T0305910

Mathematical Modeling using Fuzzy Logic

Mathematical Modeling using Fuzzy Logic
Applications to Sustainability

Abhijit Pandit

CRC Press
Taylor & Francis Group
Boca Raton London New York

CRC Press is an imprint of the
Taylor & Francis Group, an **informa** business

A CHAPMAN & HALL BOOK

First Edition published 2021
by CRC Press
6000 Broken Sound Parkway NW, Suite 300, Boca Raton, FL 33487-2742

and by CRC Press
2 Park Square, Milton Park, Abingdon, Oxon, OX14 4RN

© 2021 Taylor & Francis Group, LLC

CRC Press is an imprint of Taylor & Francis Group, LLC

ISBN: 978-1-138-39048-5 (hbk)
ISBN: 978-0-367-76792-1 (pbk)
ISBN: 978-0-429-42334-5 (ebk)

Typeset in Palatino
by codeMantra

Contents

Preface

Fuzzy logic provides a unique method for approximate reasoning in the imperfect world. This text would serve as a bridge between the principles of fuzzy logic through an application-focused approach to selected topics in Engineering and Management. The many examples point to the richer solutions obtained through fuzzy logic and to the possibilities of much wider applications.

There are relatively very few texts available at present in fuzzy logic applications. The style and content of this text is complementary to those already available. New areas of application, such as application of fuzzy logic in modeling of sustainability, are presented in a graded approach in which the underlying concepts are first described. The text is broadly divided into two parts: the first part treats processes, materials, and system applications related to fuzzy logic, and then the second part delves into the modeling of sustainability with the help of fuzzy logic.

Acknowledgments

I am thankful to almighty God, who blessed me with courage, determination, and health to complete this book.

I am very thankful to the Founder-President, Chancellor, Vice-Chancellor, Pro Vice-Chancellor, Senior Vice President, Vice President, Head of Institutions, Head of Departments, and colleagues of Amity University, Kolkata, India, for being my pillars of strength through providing constant motivation and inspiration to complete this manuscript on time.

I would also like to express my sincere gratitude to all those people who have directly or indirectly given me their precious support and valuable suggestions while pursuing this research work. I am indebted to all team members of CRC Press/Taylor & Francis Group who have constantly rendered me valuable guidance as and when required.

Last but not least, I convey my special thanks to my family members for their affection, continuous emotional support, and encouragement that inspired me throughout my mission.

Dr. Abhijit Pandit
Asst. Prof. (Grade II)
Amity University
Kolkata, India

Author

Abhijit Pandit is presently working as Assistant Professor at Amity University, Kolkata. He has more than 14 years of full-time teaching experience in reputed institutions. He has received his PhD from the University of Calcutta, MBA from MAKAUT, and M.Sc. and B.Sc. (H) from University of Calcutta. Moreover, he is a member of Indian Management Association (MIMA), lifetime member of Operational Research Society of India (ORSI), International Association of Engineers (IAENG), Society of Operations Research, and Calcutta Mathematical Society. To date, he has published 18 research papers in reputed journals, and 4 full books and 2 book chapters through international publishers. He has presented research papers in 23 conferences and also become keynote speaker in 2 such conferences. His areas of interest include quantitative techniques, fuzzy mathematics, marketing, consumer psychology, operations research, service marketing, and healthcare research. Apart from teaching and research, he is interested in music, social service, sports, etc.

1

Introduction

1.1 Rule-Based Fuzzy Logic Systems

Fuzzy logic (FL) is a type of logic that contains more and more than true or false values. It deals with situations where you can't reciprocate for yes/no (true/false) answers. In FL, propositions are expressed to a degree of truth or falsehood. In other words, FL uses a continuous range of truth values at intervals [0, 1] that are not simply true or false values. In FL, you can break both basic laws of classical logic. So, as Zadeh's Hamlet have said, "To some extent, it's a mystery." FL is a special case and contains a classic double-value logic.

According to the Encyclopedia Britannica, "Logic is the study of propositions and is used in arguments." This is the same as in Webster's list of English words: "Logic is the science of formal reasoning that uses the principles of valid reasoning," and "Logic has its main purpose. It can be applied to uncover the principles on which all valid reasoning depends and to test the justification of any conclusions drawn from the premise." Although multi-valued logic exists, we are most familiar with two-valued (double-valued) logic where the proposition is true or false. This kind of logic is called "clear logic."

Traditional (sometimes written in Western) logic was first systematized in Athens, where Aristotle propounded thousands of years ago. There are two basic laws of classical logic:

Excluded middle law: The set and its complement must contain the universe of discourse.

Law of contradiction: Elements can be sets or complementary elements. You can't have both at the same time.

These two laws sound similar, but the law of contradiction simultaneously prohibits facts that are not true, while the excluded intermediate law prohibits anything other than true or not. Shakespeare's Hamlet exemplified the law of contradiction by saying, "It's a matter of existence or not."

FL led to a new tracery for problem solving. This tracery treats inputs non-linearly and is based on rules that are logical propositions. You can extract the rules from the experts and then quantify them using FL's math you will learn in this course. This forms the tracery of fuzzy logic system (FLS), or FL's math gives the FLS the structure of a tracery from which we can tweak and use appropriate parameters to solve a problem. This helps in solving problems using neural networks (NNs). NNs' tracery is a time superiority that cannot be supported, and its parameters are tuned for troubleshooting. By combining the two approaches, you can learn a tracery that can be based on a combination of linguistic and numeric information. Both approaches play an important role in solving the problem.

> An assertion that examines only in the correct case loses all gravity if the underlying assumptions change slightly, whereas an inaccurate paper can be stable even with small perturbations of the underlying axioms.
>
> *Schwartz (1962)*

> All traditional logic habitually assumes that the correct symbols are used. Therefore, it is not possible in this earthly life, only in the imagined Godhead.
>
> *Russell (1923)*

> As the complexity of the system increases, our value in making accurate, yet important statements closer to policy, diminishes until accuracy and importance (or relevance) reaches the threshold for scrutinizing the nature of the cross-sectional area.
>
> *Zadeh (1973)*

This is classified as a principle of incompatibility.

> As we enter the information age, human knowledge is becoming more and more important. We need a theory to systematically formulate human knowledge and apply it to engineering systems along with other information such as mathematical models and sensory measurements.
>
> *L.-X. Wang (1999)*

1.2 A New Direction for FLSs

Rule-based FLS is built on the basis of logic's IF–THEN proposal, while NNs are built on simple biological models of neurons. Just as today's NN is different from biological neurons, today's rule-based FLS is different from propositional logic.

Today, fuzzy and nerve are a combination. A fuzzy NN is an NN that uses FL in some way. For example, NN's weights can be modeled as fuzzy sets. The neural fuzzy system is an FLS that uses the NN concept in some way protected by FLS.

1.3 New Concepts and Their Historical Background

In literature, it is "a processing method of describing scenes, characters, and emotions through the impression of the writer or character, not through strict objective details" (New Webster's English word list, Delair Publ. Co., 1981).

Lotfi Zadeh is the founder of FL. His first treatise on fuzzy sets came out in 1965, though he began to formulate ideas at least 4 years before. The fuzzy set encountered a massive resistance in the West due to the negative meaning associated with the word "fuzzy." Interestingly, Chinese and Japanese researchers have put a lot of effort into fuzzy sets and FL. A popular suggestion for this is that "ambiguous" goes very well with Eastern philosophy and religion (e.g. praise of yin and yang). However, until the early 1970s, FL was a theory of finding applications. "Fuzzy" does not evoke a vision of scientific or mathematical rigor. In the decades after 1965, although a relatively small number of people supported Zadeh, they learned the rigorous mathematical foundations of fuzzy sets and FL. Then, a significant transient occurred in 1975 as Mamdani and Assilian showed how to rent a nonlinear dynamic system using rule-based FL. This was relatively easy and was a quick way to diamond the rental system. Other applications of rule-based FL have started to appear. Two notable things in Japan are the subway system control and water treatment system lease in Sendai city. For example, commercial products have begun to appear such as fuzzy shower, fuzzy washing machine, and fuzzy rice cooker. In Japan, the word "fuzzy" implies "intelligent" and M. Sugeno an award in 1990. It was noted that the Western industry had to make big money, and in the decade of the 1990s, FL achieved the upper tier of fame. IEEE founded IEEE Transactions on Fuzzy Systems and established IEEE Conference and Fuzzy Systems (FUZZ). There are many other journals dealing with fuzzy systems (e.g. *Fuzzy Sets and Systems*). Moreover, there are many workshops and conferences dedicated exclusively to or including sessions on fuzzy technology. In 1995, the IEEE awarded Zadeh a medal of honor equivalent to the Nobel Prize. FL is now widely used in many industries and fields to solve practical problems and is still a topic of intensive research by scholars around the world. Although many applications have been found for FL, we are using a rule-based system that has proven its greatest efficiency as a powerful diamond methodology. This is what this rule-based fuzzy logic system (FLS) means.

If you are interested in the less impressive history of FL, you can check out the books by Wang (1999). One of FL's more important histories is provided in a recent textbook by Yen and Langari (1999, pp. 3–18).

1.4 Fundamental Diamond Requirement

Fuzzy inference systems (FISs) map vectors from input space to output space. Several other methods, including NNs, mathematical functions, and traditional tenancy systems, can also perform similar mappings. If a simpler solution comes up for a given problem, you should try it first. You can use the voluntary method instead of FLS. However, one of the many advantages of FIS is that it is flexible, it is worth modeling nonlinear functions for misdirected accuracy layers, and it is based on rules that can be specified in natural language and can be built. From their expertise, they are tolerant of inaccurate data. FL techniques can be used to complement other techniques such as NNs or genetic algorithms. To implement the FIS, you first need to determine the inputs, outputs, each domain, and the rules of fuzzy inference. Mapping rules can be obtained from numeric data or expertise. You also need to determine the input and output membership features, the overlap between these features, the implications and teaming methods, and how to defer to. The designation of these parameters allows you to make a variety of FIS diamonds. You can choose any method, algorithm, and programming environment to implement FIS. Many software and hardware tools today are implemented in designing and implementing fuzzy logic systems (FLS). Most software tools provide extensive debugging and optimization capabilities as well as a graphical user interface (GUI) environment that makes FLS simple and easy. One such package is the MATLAB® Fuzzy Logic Toolbox, which relies heavily on GUI tools to provide an environment for FIS design, analysis, and implementation. The example shown in the previous section can be solved using this toolbox. MATLAB provides five basic GUI tools for building, editing, and observing FIS.

1.5 The Flow of Uncertainties

Uncertainty is the main reason humans make their own choices. Many actions and strategies are designed to deal with or reduce uncertainty to make the life of the researcher easier. But in academia, there are no phenomenal theories around the world. Researchers are still struggling to understand globally.

There are some theories you can follow to make decisions in uncertain situations. Among them are probability theory, fuzzy theory, and trace theory. Visualization problems are synthesized with the help of these theories. FL and fuzzy theory can be considered a recent urge and have been unromantic in many fields for different types of visualization problems.

Uncertainty has been studied in many fields over the past few centuries. The discussion of uncertainty can be marvelous in the field you are pursuing, but it is not limited to this. I would like to summarize some important facts and research struggling to discuss uncertainty: visualization science, tense intelligence, legal facts, economics, medicine, organizational systems theory, psychology, and physics.

Often the author starts with a definition of a word list of uncertainty. We all use this term in our daily life. To annuity our subtitle compact, we provide these definitions. However, some words related to uncertainty, such as inaccuracy, ambiguity, and ambiguity, are important. Surprisingly, the information usually comes with some words that don't have an exact definition. "Information," "Low," "High," "Large," "Fast," and "Slow" are some examples of misleading reasoning in the creation of human visualizations. Humans understand it and use it to communicate strongly. This word comes from a specific field of study, Computing With Words (CWW). Their imprecise meaning is also considered a starting point for FL. Even if it is inaccurate, it helps us communicate and decide which flight.

Researchers have tried to identify the different types and dimensions of uncertainty. Among them, Smithson, Smets, Bosc and Prade, Klir and Yuan, Walker and Parsons may be mentioned.

Uncertainty has to do with the truth or certainty of the information. As Bosc and Prade suggest, uncertainty that arises from the "lack of information" closely related to probabilistic theory advocates who evaluate probability as a lack of knowledge. Dubois and Prade distinguish between the concepts of inaccuracy and uncertainty in a way that relates them to the content of information value. Like them, the concept of fuzzy is a qualifier for an item of information. Klir and Yuan identify three important types of uncertainty. These are non-specificity, conflict, and ambiguity. Like Walker, the Uncertainty Theory has three dimensions in the investigation of legal facts. These are the linguistic, logical, and causal dimensions. Walker classified uncertainty into six types of scientific evidence close to common causes. These are conceptual uncertainty, measurement uncertainty, numbering uncertainty, sampling uncertainty, mathematical modeling uncertainty, and causal uncertainty. In Smithson's classification, uncertainty is part of imperfection, the product of error. It can be seen as a certain type of ambiguity that spreads to the researcher. Smets sees uncertainty as an important part of ignorance. His classification puts the concept of fuzzy under the data without the erroneous portion of uncertainty.

Uncertainty is a cognitive process. It is advisable to reward different perspectives in your chosen field. Knight and Keynes, for example, considered uncertainty as "simply we don't know." Uncertainty is considered "lack of knowledge," "prejudice," and "psychological perception," at least in many areas, such as in Keynes and Knight's demonstrations. Uncertainty looks like this in psychology: (1) psychological perceptions that cause fear and (2) motivations for communication.

Information and knowledge have ripple effects in the real world. Quantitative measures like probability can turn out to be insufficient or misleading in many cases. Humans seem to try to overcome these difficulties by using heuristics as the first tool for reasoning under uncertainty. It doesn't help and is a quick and dirty way to deal with uncertainty. You need a formal system that can handle uncertainty to ensure that the information is used strongly. To do this, you can find three widely studied systems in the literature: probability, probability, and trace theory.

Every visualization we create is the result of a series of negotiations working to reduce uncertainty, but it is part of the source of the uncertainty side. This is the main element of game theory. Discussion of the risk game leads to an idea of the expected usefulness. A "reasonable" way to create visualizations based on Expected Utility Theory (EUT) is to maximize expected utility. However, incomplete knowledge makes life difficult for EUT supporters. Friedman and Savage proposed the axiom of the subjective EUT. To demonstrate the systematic deviation from the EUT, Allais has released "Allais Paradox." Thus, the Unexpected Utility Theory (UEUT) emerged. In the late 1960s, visualization of ignorance was revived. Instead of normative theories, researchers began to squint with increasingly descriptive decision theories. Several important contributions have been made to the visualization under uncertainty (sometimes marked as tabs under "Ignorance" and "Danger"). Among the contributors are Kahneman and Tversky, Tversky and Kahneman, Machina, Fishburn, Kahneman et al., Lichtenstein, and Slovic.

1.6 Existing Literature on Type 2 Fuzzy Sets

Type 2 fuzzy sets allow you to model and minimize the impact of uncertainty in rule-based FLS. There are (at least) four sources of uncertainty in type 1 FLS. (1) The meaning of words used in guardians and the consequences of the rules can be uncertain (words mean different things to different people). (2) Results can explicitly have a histogram of relevant values when extracting knowledge from a group of experts who disagree. (3) The measurement that activates the type 1 FLS is noisy and can be uncertain. (4) The data used to adjust the parameters of the type 1 FLS may be noisier. All of these uncertainties translate to uncertainties that are close to the fuzzy set membership function. Given the fact that the type 1 fuzzy set has a completely sharp membership function, it is not wise to model this uncertainty yourself. On the other hand, type 2 fuzzy sets are worldly for modeling such uncertainty, given that the membership function itself is ambiguous. The membership function of a type 1 fuzzy set is two-dimensional, whereas the membership function of a type 2 fuzzy set is three-dimensional. It's a new three-dimensional set of type 2 fuzzy, which provides your own rules of slack level to directly model the uncertainty.

Even in situations where these difficulties were scattered, type 2 fuzzy sets and FLS were used once (this list is presented in alphabetical order depending on the application): classification of coded video streams, support for co-channel interference from nonlinear time-varying contact channels, connections to ticket tenancy, mobile robot tenancy, visualization production, equalization of nonlinear fading channels, knowledge extraction from surveys, time series prediction, functional propensity, language member class learning, radiographic image pre-processing, relational database, fuzzy relationship solving equation, and transmission schedule.

(1) When the data generation system is known to change with time, the mathematical explanation of the time fluctuation is unknown (for example, as in mobile communication). (2) The measurement noise is abnormal, and the mathematical explanation of the abnormality is unknown. (3) The characteristic of pattern recognition used has an abnormal statistical characteristic, and the mathematical explanation of the abnormality is unknown. (4) Knowledge is mined from a group of experts using questionnaires containing uncertain words. (5) Linguistic terms with areas that cannot be measured are used.

Type 2 fuzzy sets are difficult to understand and use, considering the following: (1) It is very difficult to draw due to the three-dimensional nature of the type 2 fuzzy set. (2) There is no mere confusion of well-defined terms so that we can communicate well close to type 2 fuzzy sets and have them close mathematically exactly (the terms exist but are not precisely defined). (3) The derivation of all formulas for combining, crossing, and complementing the type 2 fuzzy sets relies on the use of Zadeh's extension principle, which in itself is a difficult concept (especially for those new to FL) and somewhat ad hoc. Using it to derive things can be problematic. (4) Using a type 2 purge set becomes more and more computationally complex than using a type 1 purge set. Difficulty is the price you have to pay to achieve your greatest performance in a situation where uncertainty is scattered, consistent with using probability rather than determinism.

1.7 Coverage

FL is a formal mathematical flow for modeling decentralized systems that have been successfully used in many tenancy systems. Membership features and language descriptions in the form of rules make up the model. Rules are created a priori from data through expertise or system identification methods. The input membership function is based on an estimate of the ambiguity of the descriptors used. The output membership function can be set initially, but it can be modified for controller adjustment. Once these are defined, the operating procedure for the calculation is well established. Measurement data is converted into membership through a fuzzification process. The rule is non-romantic, which uses typed actions to generate members of the output set. Finally, these are combined via defuzzification to compensate for the final tenancy output. Fuzzy value for inaccuracies due to the use of linguistic terms. The use of FL in modulation chemistry applications involves the inaccuracies of the measurement method in the numbering of the final output. This reasonableness can enable the use of noisy measurement systems.

1.8 Applicability Outside of Rule-Based FLS

FLS can be specified as a nonlinear mapping of input data sets to scalar output data. FLS consists of four main parts: fuzzy fire, rules, inference engine, and defuzzifier.

The process of FL is described in Algorithm 1.1. First, we collect a well-written input data set and transform it into a fuzzy set using fuzzy language variables, fuzzy language terms, and membership functions. This step is called fuzzification. After that, inference is made based on a set of rules. Finally, the resulting fuzzy output is mapped to a well-completed output using the membership function in the deepening phase.

To illustrate the use of FLS, consider an air-conditioning system controlled by FLS. The system adjusts the room temperature equal to the room's current temperature and target value. The purge engine periodically compares the room temperature to the target temperature and generates a record for heating or burning the room.

Algorithm 1.1: Fuzzy Logic Algorithm

1. Definition of language variables and terms (initialization)
2. Membership function establishment (initialization)
3. Rule knowledge base (initialization) configuration
4. Convert well-written input data to fuzzy values using the membership function (fuzzy)
5. Evaluate the rules in rule wiring (inference).
6. Combining the results of each rule (inference)
7. Convert output data to non-fuzzy values (differentiation)

1.9 Computation

CWW (Zadeh, 1996) deals with words and propositions from a natural language as the main objects of computation; for example, "small," "big," "expensive," "quite possible," or plane increasingly ramified sentences such as "tomorrow will be cloudy but not very cold." The main inspiration of CWW is the human worthiness of performing several variegated tasks (walk in the street, play football, ride a bicycle, understand a conversation, making a decision, etc.) without needing an explicit use of any measurements or computations. This sufficiency is sustained by the brain's worthiness to manipulate variegated perceptions (usually imprecise, uncertain, or partial perceptions), which plays a key role in human recognition, visualization, and execution processes.

CWW is a methodology that contrasts with the usual sense of computing, that is, manipulating numbers and symbols. Therefore, CWW provides a methodology that can bring closer the gap between human's smart-ass mechanisms and the machine's processes to solve problems providing computers with tools to deal with imprecision, uncertainty, and partial truth.

Decision making (Triantaphyllou, 2000) can be seen as the final outcome of some mental processes that lead to the selection of a volitional one among several variegated ones. For example, typical visualization-making problems are to segregate the weightier car to buy, deciding which candidate is increasingly suitable for an unrepealable position in a firm, or choosing the most towardly gown for a meeting. It is interesting to note that visualization making is an inherent human worthiness which is not necessarily rationally guided, can be based on explicit or tacit assumptions, and does not need precise and well-constructed measurements and information well-nigh the set of feasible alternatives. This fact has led many authors to wield fuzzy sets theory (Zadeh, 1965) to model the uncertainty and vagueness in visualization processes.

In recent years, many researchers have seen CWW as a very interesting methodology to be unromantic in visualization making (Herrera & Herrera-Viedma 2000). As it allows to model perceptions and preferences in an increasingly human-like style, it can provide computers some of the needed tools, if not to fully simulate human visualization making, to develop ramified visualization support systems to ease the visualization makers to reach a solution.

1.10 Primer on Fuzzy Sets

1.10.1 Fuzzy Sets

A crisp set is set specified using a foible function that assigns a value of either 0 or 1 to each element of the universe, thereby discriminating between members and non-members of the crisp set under consideration. In the context of fuzzy sets theory, we often refer to well-done sets as "classical" or "ordinary" sets. A conventional set for which an element is either a member of the set or not is a crisp set. In a crisp set, elements have a Boolean state that implies either membership exists or not.

1.10.2 From Fuzzy Sets to Crisp Sets

Defuzzification is the process of producing a quantifiable result in crisp logic, given fuzzy sets and respective membership degrees. It is the process that maps a fuzzy set to a crisp set. It is typically needed in fuzzy tenancy systems. These will have a number of rules that transform a number of variables into a fuzzy result, that is, the result is described in terms of membership in fuzzy sets. For example, rules designed to decide how much pressure to wield might result in "Decrease Pressure (15%), Maintain Pressure (34%), Increase Pressure (72%)." Defuzzification is interpreting the membership degrees of the fuzzy sets into a specific visualization or real value.

The simplest but least useful defuzzification method is to segregate the set with the highest membership, in this case, "Increase Pressure," since it has a 72% membership, and ignore the others, and convert this 72% to some number. The problem with this tideway is that it loses information. The rules that tabbed for decreasing or maintaining pressure might as well have not been there in this case.

A worldwide and useful defuzzification technique is partway of gravity. First, the results of the rules must be aggregated together in some way. The most typical fuzzy set membership function has the graph of a triangle. Now, if this triangle were to be cut in a straight horizontal line somewhere between the top and the bottom, and the top portion were to be removed, the remaining portion forms a trapezoid. The first step of defuzzification typically "chops off" parts of the graphs to form trapezoids (or other shapes if the initial shapes were not triangles). For example, if the output has "Decrease Pressure (15%)," then this triangle will be cut 15% the way up from the bottom. In the most worldwide technique, all of these trapezoids are then superimposed one upon another, forming a single geometric shape. Then, the centroid of this shape, tabbed the fuzzy centroid, is calculated. The x coordinate of the centroid is the defuzzified value.

There are many variegated methods of defuzzification available, including the following:

- AI (adaptive integration)
- BADD (basic defuzzification distributions)
- BOA (bisector of area)
- COA (center of area)
- COG (center of gravity)

- EQM (extended quality method)
- FCD (fuzzy clustering defuzzification)
- FM (fuzzy mean)
- FOM (first of maximum)
- GLSD (generalized level set defuzzification)
- IV (influence value)
- LOM (last of maximum)
- MeOM (mean of maxima)
- MOM (middle of maximum)
- QM (quality method)
- SLIDE (semi-linear defuzzification)
- WFM (weighted fuzzy mean)

The maxima methods are good candidates for fuzzy reasoning systems. The distribution methods and the zone methods walk out the property of continuity that makes them suitable for fuzzy controllers.

1.10.3 Linguistic Variables

While variables in mathematics usually take numerical values, in FL applications, non-numeric values are often used to facilitate the expression of rules and facts.

A linguistic variable such as age may winnow values such as young and its antonym old. Considering that natural languages do not unchangingly contain unbearable value terms to express a fuzzy value scale, it is a worldwide practice to modify linguistic values with adjectives or adverbs. For example, we can use the hedges rather and somewhat to construct the spare values rather old or somewhat young.

Fuzzification operations can map mathematical input values into fuzzy membership functions. And the opposite defuzzifying operations can be used to map a fuzzy output membership functions into a "crisp" output value that can be then used for visualization or tenancy purposes.

In practice, when the universal set X is a continuous space (the real line R or its subset), we usually partition X into several fuzzy sets whose Membership Functions (MFs) imbricate X in an increasingly or less uniform manner. These fuzzy sets, which usually siphon names that conform to adjectives seeming in our daily linguistic usage, such as "large," "medium," or "small," are tabbed linguistic values or linguistic labels. Thus, the universal set X is often a tabbed linguistic variable. Formal definitions of linguistic variables and values are given in this section. Here is a simple example.

Suppose that X = "age." Then, we can pinpoint fuzzy sets "young," "middle aged," and "old" that are characterized by MFs $\mu young(x)$, $\mu middleaged(x)$, and μold(x), respectively. Just as a variable can have various values, a linguistic variable "Age" can have variegated linguistic values, such as "young," "middle aged," and "old" in this case. If "age" assumes the value of "young," then we have the expression "age is young," and so withal for the other values. A fuzzy set is unique specified by its membership function.

1.10.4 Membership Functions

Zadeh introduced the term "fuzzy logic" in his seminal work "Fuzzy sets," which described the mathematics of fuzzy set theory (1965). Plato laid the foundation for what would wilt FL, indicating that there was a third region vastitude True and False. It was Lukasiewicz who first proposed a systematic volitional to the bi-valued logic of Aristotle. The third value Lukasiewicz proposed can be weightier translated as "possible," and he prescribed it a numeric value between True and False. Later he explored four- and five-valued logics, and then he supposed that, in principle, there was nothing to prevent the derivation of infinite-valued logic. FL provides the opportunity for modeling conditions that are inherently imprecisely defined. Fuzzy techniques in the form of injudicious reasoning provide visualization support and expert systems with powerful reasoning capabilities. The permissiveness of fuzziness in the human thought process suggests that much of the logic overdue thought processing is not traditional two-valued logic or plane multi-valued logic, but logic with fuzzy truths, fuzzy connectivity, and fuzzy rules of inference. A fuzzy set is an extension of a crisp set. Crisp sets indulge only full membership or no membership at all, whereas fuzzy sets indulge partial membership. In a crisp set, membership or non-membership of element x in set A is described by a characteristic function $\propto_A (x)$, where $\propto_A (x) =$ 1 if $x \in A$ and $\propto_A (x) = 0$ if $x \notin A$. Fuzzy set theory extends this concept by defining partial membership. A fuzzy set A on a universal set U is characterized by a membership function $\propto_A (x)$ that takes values in the interval [0, 1]. Fuzzy sets represent nous linguistic labels like slow, fast, small, large, heavy, low, medium, high, tall, etc. A given element can be a member of increasingly than one fuzzy set at a time.

A membership function is substantially a curve that defines how each point in the input space is mapped to a membership value (or stratum of membership) between 0 and 1. As an example, consider height as a fuzzy set. Let the universe of discourse be heights from 40 to 90 inches. In a crisp set, all people with a height of 72 inches or more are considered tall, and all people with a height of <72 inches are considered not tall. The curve defines the transition from not tall and shows the stratum of membership for a given height. We can proffer this concept to multiple sets. If we consider a universal set from 40 to 90 inches, then we can use three term values to describe height, namely, short, average, and tall. In practice, the terms short, average, and tall are not used in a strict sense. Instead, they imply a smooth transition.

1.10.5 Some Terminology

A fuzzy rule is a simple IF–THEN rule with a condition and a conclusion. Sample fuzzy rules for the air conditioner system are as follows:

1. IF (temperature is warm OR too-cold) AND (target is warm), THEN action is heat.
2. IF (temperature is hot OR too-hot) AND (target is warm), THEN action is cool.
3. IF (temperature is warm) AND (target is warm), THEN action is no-change.

1.10.6 Set-Theoretic Operations on Crisp Sets

Crisp sets consist of mishmash of well-defined objects. It is well defined in the sense that an object is applied to a set or not. Here are some of the most important tasks in a crisp set.

Union
The union of two sets A and B is a set that contains elements of both sets A and B. This is indicated by (A U B).

Intersection
The intersection of both sets A and B is a set of amazingly contained elements in both set A and set B. This is indicated by (A ∩ B).

Complement
The complement of set A is the set of all elements in the universal set "E" but not in set A. The complement of set A is denoted by A^c.

Set Difference
The difference between sets A and B is in A as A-B, but only the set of all elements in B or all other elements in E. This is indicated by (A-B).

1.10.7 Set-Theoretic Operations for Fuzzy Sets

Given "X" to be universe of discourse, A and B are two fuzzy sets with membership function $\mu A(x)$ and $\mu B(x)$; then,

Union
The union of two fuzzy sets A and B is a new fuzzy set A ∪ B.

Intersection
Intersection of fuzzy sets A and B is a new fuzzy set A ∩ B.

Equality
Two fuzzy sets A and B are said to be equal, i.e. A=B if and only if $\mu A(x) = \mu B(x)$ which ways their membership values must be equal.

1.10.8 Crisp Relations and Compositions on the Same Product Space

A crisp relation R from a set A to a set B assigns to each ordered pair exactly one of the pursuit statements: (i) "a is related to b," or (ii) "a is not related to b."

The Cartesian product A×B is the set of all possible combinations of the items of A and B. For example, when

$$A = \{a_1, a_2, a_3\} \text{ and } B = \{b_1, b_2\}$$

$$A \times B = \{(a_1, b_1), (a_1, b_2), (a_2, b_1), (a_2, b_2), (a_3, b_1), (a_3, b_2)\}$$

1.10.9 Relations and Compositions

A crisp relation R from a set A to a set B assigns to each ordered pair exactly one of the pursuit statements: (i) "a is related to b," or (ii) "a is not related to b."

The Cartesian product A×B is the set of all possible combinations of the items of A and B. For example, when

$$A = \{a_1, a_2, a_3\} \text{ and } B = \{b_1, b_2\}$$

$$A \times B = \{(a_1, b_1), (a_1, b_2), (a_2, b_1), (a_2, b_2), (a_3, b_1), (a_3, b_2)\}$$

Fuzzy relations map elements of one universe, say U, to those of flipside universe, say V, through the Cartesian product of the two universes. However, the "strength" of the relation between ordered pairs of the two universes is measured with a membership function expressing various "degrees" of strength of the relation on the unit interval [0,1].

As an example, a fuzzy relation "Friend" describes the stratum of friendship between two persons (in unrelatedness to either stuff friend or not stuff friend in classical relation!)

Fuzzy relation "Similarity" $U = V = \{1, ..., 8\}$.

	1	2	3	4	5	6	7	8
1	1	0.5	0	0	0	0	0	0
2	0.5	1	0.5	0	0	0	0	0
3	0	0.5	1	0.5	0	0	0	0
4	0	0	0.5	1	0.5	0	0	0
5	0	0	0	0.5	1	0.5	0	0
6	0	0	0	0	0.5	1	0.5	0
7	0	0	0	0	0	0.5	1	0.5
8	0	0	0	0	0	0	0.5	1

Let A be a fuzzy set specified on a universe of three discrete temperatures, $X = \{x_1, x_2, x_3\}$, and B be a fuzzy set specified on a universe of two discrete pressures, $Y = \{y_1, y_2\}$. Fuzzy set A represents the "ambient" temperature, and fuzzy set B the "near-optimum" pressure for a unrepealable heat exchanger, and the Cartesian product represents the conditions (temperature–pressure pairs) of the exchanger that are associated with "efficient" operations. For example, let

$$A = 0.2/x_1 \quad 0.5/x_2 \quad 1/x_3 \quad \text{and} \quad B = 0.3/y_1 \quad 0.9/y_2$$

	y_1	y_2
Then, $A \times B = R = x_1$	0.2	0.2
$= x_2$	0.3	0.5
$= x_3$	0.3	0.9

The fuzzy relation is one kind of fuzzy sets. Therefore, we can wield operations of fuzzy set to the relations (e.g. Union, Intersection, Complement).

A fuzzy relation R is specified on sets A and B, and a flipside fuzzy relation S is specified on sets B and C:

$$\text{That is, } R \subseteq A \times B, S \subseteq B \times C$$

The sonnet $S \bullet R = SR$ of the two relations R and S expresses the relation from A to C.

This sonnet is specified by an inner product. The inner product is similar to an ordinary matrix (dot) product, except that multiplication is replaced by minimum, while summation is replaced by maximum. Thus, this sonnet is specified by the pursuit:

$$\mu S \cdot R(a,c) = \max\left[\min\left(\mu R(a,b), \mu S(b,c)\right)\right]$$

1.10.10 Hedges

In the 1970s, Zadeh introduced and mastered the theory of detrimental reasoning based on the concept of linguistic variables and FL. Informally, by linguistic variable, we designate a variable whose value is a word in natural or tense language as an intermediate point. For example, Age is a language variable with linguistic values such as young, old, very young, very old, increasingly younger, not very young, not very old, etc. Language from primary terms (e.g. young and old in a sample of linguistic variable age) by values various language hedges (e.g. very, increasingly or less) and connections (e.g. and, or not) as well known creation of variables. In terms of Zadeh's FL, the truth value is verbal. For example, "True," "Very true," "False," "False," "Possibly False," etc. can be used to express the value of the language variable Truth. In this sense, malicious reasoning (also tab-separated fuzzy reasoning) is mostly qualitative rather than quantitative in nature, and upon closer examination, everything falls outside the sphere of applicability of classical logic. The main purpose of detrimental reasoning theory is to mimic human verbal reasoning, especially when describing the behavior of human-centric systems. In the same way as Zadeh's rule for qualification for truth, propositions such as "Lucia is very young" are considered to be semantically equivalent to the proposition "Lucia is young." This semantic equivalence relationship plays an important role in false reasoning. In a fuzzy set–based approach to fuzzy inference, primary linguistic truth values, such as true and false, are stipulated to correspond to the fuzzy set specified in an interval (0, 1) designed to interpret the meaning of that primary term. Then, the complex language truth value is computed using a tracking procedure:

- For example, more or less linguistic hedges are specified as unary operators in fuzzy sets.
- Logical concatenations such as "and," "or," "not," and "if... then" are specified by operators such as "t-norm," "t-conorm," "negation," and "implication," respectively. As is well known, one of the inherent problems of fuzzy inference models is linguistic approximation, that is, how to name the resulting fuzzy set of the inference process in linguistic terms. This depends on the shape of the resulting fuzzy set with respect to the main fuzzy set and operator. Based on the two characteristics of linguistic variables introduced by Zadeh (i.e. the context-independent meaning of linguistic hedge and linkage, the universality of the domain of language) and the meaning of linguistic hedge in natural language, Nguyen and Wechler proposed a flow of trigonometry. Moreover, the assistant gives us the possibility to introduce verbal reasoning methods that can directly deal with linguistic terms and thus avoid the problem of linguistic approximation.

1.10.11 Expansion Principle

An extension principle learned by Zadeh is a function for fuzzy sets. Thus, it generalizes the global point-to-point mapping of the function to the mapping between fuzzy sets.

1.11 FL Primer

FL is similar to human decision-making methodology. It deals with vague and inaccurate information. This is an oversimplification of the real problem and is based on the degree of truth rather than 1/0 like normal true/false or Boolean logic.

Fuzzy logic is not a fuzzy logic, but a logic used to describe fuzzy. FL was introduced in 1965 by Lofti A. Zadeh in his research paper "Fuzzy sets." He is considered the father of FL.

1.11.1 Crisp Logic

The traditional flow of knowledge representation (clear logic) does not provide a way to interpret inaccurate and non-categorical data, since its function is based on first-order logic and classical probability theory. Conversely, it cannot deal with the expression of human intelligence.

Now let's understand the clear logic with an example. We have to find a pun for the question "Does she have a pen?" The pun in the presented question is yes or no depending on the situation. If yes is qualified with the value 1 and no is qualified with 0, the result of the statement can be 0 or 1. Therefore, any logic that requires binary (0/1) type of processing is known as well-done logic.

1.11.2 From Crisp Logic to FL

Fuzzy set theory is intended to introduce ambiguity and inaccuracies to model human smartness in tense intelligence, and the importance of such a theory is increasing day by day in the field of expert systems. However, the clear set theory was very constructive as an early concept for modeling digital and expert systems operating on binary logic. The fuzzy set is not swayed by an uncertain boundary, and there is uncertainty around the set boundary. The sharp set, on the other hand, is specified as a well-finished boundary and contains the exact location of the set boundary. Fuzzy set elements are allowed partially uncomplained by the set (indicating the progressive membership degree). Conversely, sharp set elements can have full membership or non-membership. There are several applications of fuzzy set theory that perform well and are both aimed at the details of an efficient expert system. Fuzzy sets follow an infinite-value logic, while well-done sets are based on a double-value logic.

Fuzzification is the process of converting well-completed input values into fuzzy values that are performed using information from a knowledge base. Although you can see different types of curves in the literature, Gaussian, triangular, and trapezoidal MF is most surprisingly used in the fuzzification process. This type of MF can be implemented with an embedded controller.

MF is specified mathematically using several parameters. You can adjust these parameters or the shape of the MF to fine-tune the performance of the Fuzzy Logic Control (FLC). Fuzzification is the process of breaking down system inputs and/or outputs into one or more fuzzy sets. You can use many types of curves and tables, but membership functions in the shape of a triangle or trapezoid are the most common because they are easier to express in an embedded controller.

Consider a simplified implementation of an air-conditioning system with a temperature sensor. The temperature can be read by a microprocessor with a fuzzy algorithm that processes the output to keep the motor in a consistent speed for maintaining the room at a "good temperature." You can also interpret the vents upwards or downwards as needed. There are five purge sets for temperature: COLD, COOL, GOOD, WARM, and HOT.

The membership features of the fuzzy sets COOL and WARM are trapezoidal, the membership features of GOOD are triangular, the membership features of COLD and HOT are half triangular, and the shoulders represent the physical limits to these processes (if you are staying in low room temperature). It will be much more inconvenient in temperatures less than 8°C or more than 32°C. Diamonding such a fudge set is entirely up to the designer's wit and intuition. An input temperature of 18°C is considered COOL to the extent of 0.75 and GOOD for a strata of 0.25. To build rules for renting air conditioner motors, you can see how human experts empirically get the rules by adjusting settings to speed up or slow down the motor in vibrations over temperature. For example, if the room temperature is good, designate the motor speed medium. If the temperature is high, turn the speed control quickly, and if the room is hot, increase the speed. On the other hand, when the temperature is low, it slows down and stops when the motor is cold. What stands out for FL is the way it turns global sense and verbal explanations into computer-controlled systems. To organize this process well, you need to understand how to create rules using some logical operations.

Boolean logic operations should be extended to FL to manage the concept of partial truth (the truth value between "fully true" and "fully false"). The fuzzy nature of a sentence such as "X is LOW" can be combined with a fuzzy sentence of "Y is HIGH," and since X is LOW and Y is HIGH, normal logical operations can be given. What is the true value of this AND operation? Logical operations with fuzzy sets are performed as a membership function. There are various interpretations of FL operations, but the trace definition is very user friendly in embedded tenancy applications:

truth(X and Y)=Min(truth(X), truth(Y))

truth(X or Y)=Max(truth(X), truth(Y))

truth(not X)=1.0 −truth(X)

1.12 Remarks

The concept of FL was first presented by Zadeh, which is able to translate our qualitative knowledge into reasoning systems. Hybrid intelligent Phased

Mission System (PMS) inspired by the proposed quantum is designed with the help of FL. As far as PMS is concerned, we used FL as a problem-solving tenancy system approach. FL provides a decent way to solve location-related queries that are ambiguous and contain inaccurate input information. Also, instead of applying a mathematical model to a specific system, FL incorporates rule-based IF A AND B THEN C flows to solve position-based queries. Therefore, it incorporates rule-based tactics into the proposed PMS framework to resolve location-related queries in real time and provide confidentiality protection for end users' current location. Also in FL, the concept of a fuzzy reasoning system is used. This is basically the process of formulating a mapping from a given input to an expected output. It consists of IF–THEN rules, logical operations and membership functions. FIS can be implemented using two techniques: Mamdani type and Sugeno type techniques. The fuzzy reasoning system process consists of five steps: fuzzifying input variables, applying fuzzy operators (AND or OR), inference from predecessors (predecessors) to results, teaming results through rules, and fuzzifying. In the proposed PMS framework, we integrate the concept of FIS and FL. Location-based query forwarding and response is implemented in a conceptual way of FIS.

1.13 Exercise

1. Distinguish between fuzzification and defuzzification.
2. What is the main difference between probability and fuzzy logic?
3. What is the primary reason that fuzzy logic has wilt one of the most successful tools in developing sophisticated system?
4. Describe any one method of defuzzification.
5. Discuss various set-theoretic operations on fuzzy sets with relevant examples.

References

Herrera, Francisco, and Enrique Herrera-Viedma. "Linguistic decision analysis: Steps for solving decision problems under linguistic information." *Fuzzy Sets and Systems* 115, no. 1 (2000): 67–82.

Russell, Bertrand. "Vagueness." *The Australasian Journal of Psychology and Philosophy* 1, no. 2 (1923): 84–92.

Schwartz, Jack "The harmful effects of mathematics in science," In Ernest Nagel, Patrick Suppes, and Alfred Tarski (Eds), *Logic Methodology and Philosophy of Science*, pp. 231–235. Standford University Press, Standford, CA (1962).

Schwartz, Jack. "The pernicious influence of mathematics on science." In *Studies in Logic and the Foundations of Mathematics*, vol. 44, pp. 356–360. Elsevier, Amsterdam, Netherlands (1966).

Triantaphyllou, Evangelos. "Multi-criteria decision making methods." In: *Multi-Criteria Decision Making Methods: A Comparative Study*, pp. 5–21. Springer, Boston, MA (2000).

Wang, Liang, Reza Langari, and John Yen. "Identifying fuzzy rule-based models using orthogonal transformation and backpropagation." In *Fuzzy Theory Systems*, pp. 187–204. Academic Press, Cambridge, MA (1999).

Zadeh, Lotfi Asker. "Fuzzy sets." *Information and Control* 8, no. 3 (1965): 338–353.

Zadeh, Lotfi Asker. "The concept of a linguistic variable and its application to approximate reasoning, memorandum ERL-M411, Berkeley." (1973).

Zadeh, Lotfi Asker. "Fuzzy logic=computing with words." IEEE Transactions on Fuzzy Systems 4, no. 2 (1996): 103–111. DOI: 10.1109/91.493904.

2

Sources of Uncertainty

2.1 Uncertainty

Uncertainty has been studied in many fields over the past few centuries. The discussion of uncertainty can be marvelous in, but is not limited to, the field you are pursuing, such as visualization science, tense intelligence, legal facts, economics, medicine, organizational systems theory, psychology, and physics (McCorduck & Cfe 2004).

Uncertainty is the main reason why humans make their own choices. Visualization problems are synthesized with the help of these theories. Many actions and strategies are designed to deal with or reduce uncertainty to make the life of the visualization creator easier. But in academia, there are no phenomenal theories around the world. Researchers are still struggling to understand uncertainty globally. There are some theories you can follow to make decisions in uncertain situations. Among them are probability theory, fuzzy theory, and trace theory. Fuzzy logic and fuzzy theory can be considered a recent urge, but these have been undesirable in many fields due to different types of visualization problems.

The concept of fuzzy sets and their extensions has been understood in various ways in the literature. There are several concepts that appeal to the fuzzy set:

- Graduality: The idea that many categories (natural languages) contain truths compared to the Boolean tradition.
- Epistemic uncertainty: The idea of expressing partial or incomplete information in sets, e.g. possibility theory, and modal logic versus stochastic tradition.
- Ambiguity: The idea that there are no well-known truth conditions for the extension of natural language predicates, e.g. classical logic nightmare.

2.1.1 Uncertainty: General Discussion

Measurement results are often used to estimate a specific quantity indicated by a tab on the measurement quantity. The difference between the measurement result and the measurement value is a measurement error. Measurement results and errors may vary with each repeat of the measurement, but the measurand value (actual value) remains fixed.

Systematic effect rationalization results tend to differ from the value of the measurand by relative quantities, or tend to change in a non-random way. Usually, random and systematic effects exist in the measurement process (Viechtbauer, 2007).

The random effect rationalizes the measurement results so that they change randomly as measurements are repeated. Measurement errors can occur due to arbitrary and systematic effects in the measurement process.

Also, measurement errors can lead to fake results such as wrong results due to human errors and malfunction of the device. Mistakes and other spurious errors are of no value in the statistical evaluation of measurement uncertainty. Where possible, good laboratory practices should be used to avoid or at least detect and correct through quality assurance and quality control.

2.1.2 Uncertainty at FLS

Graduality, epistemological uncertainty, and ambiguity seem to interact closely with one side within fuzzy set theory. Because fuzzy sets expand the concept of the set, it may be worthwhile for epistemological uncertainty. Because quantification is often a matter of degree, epistemological uncertainty is gradual. Sometimes, the membership feature can be of value for the lesser-known Purius and can be seen as a modeling ambiguity.

According to Keefe and Smith, there are three characteristics of ambiguity in natural language in the presence of uncertain cases: violation of the ruled-out intermediate law, difficulty in visualizing the truth or falsehood of ambiguous sentences, and the use of planar specimens for accurate explanations (Keefe 2000). But is ambiguity a flaw in natural language (not for devising formal proofs), or is it quality (tolerance to errors, flexibility for communication purposes)? The main debate seems to be against those whose ambiguous predicates call for an unknown standard extension and those who reject the existence of visualization thresholds.

As Zadeh (1978) writes,

> fuzzy and ambiguous terms are used interchangeably in the literature, but in reality, there are significant differences between the two. Specifically, the proposition p is fuzzy if it contains words that are the labels of the fuzzy set. p is ambiguous and ambiguous if it is insufficiently specific for a particular purpose. For example, 'Bob will be when in a few minutes' is ambiguous, whereas 'Bob will be when in a few

minutes' is ambiguous when the infrastructure for the decision is not enough. So the ambiguity of a proposition is a loophole that depends on decision making, but ambiguity is not.

For Zadeh, ambiguity is gradual and ambiguity is ambiguity without specificity.

If P is the ambiguous property, the truth or falsehood of P may be personal for some objects, but not for all. The truth of P is not considered gradual. The uncertain factor is the factor that the truth or falsehood of P cannot reach.

You can claim the truth of P for an object.

- Supervaluation: For all methods of sharpening P (super-truth for Fine).
- Epistemic view: If P is known to be true (obviously true in Williamson's case), in both approaches, the truth of P is established under the same conditions. The only difference is whether P's "real" sharpener exists. The extension of P is an unknown set.

2.2 Words Mean Different Things to Different People

Remember that the dictionary can be larger if you think there are too many words in English. Our language has doubled the definition of many words.

Heteromorphs and homophones all cause problems especially when trying to use these double-edged words while avoiding spelling errors and misunderstandings. When things get so complicated, it's time to uncover the dusty old tormented language, as well as Greek suffixes and prefixes, which are very important in deciphering the roots of a word (Fry, 1988).

The difference between homophones and dimorphisms lies in the combination of roots and spellings, pronunciation, and well-expressed elements. There are also various words with two meanings, including various similarities beyond the gap between these three unshared segments.

Homophones are words that are pronounced in the same way, but with different meanings. Words with different meanings and different spellings that are pronounced in the same way are known as heterogeneous graphs. Examples of words that sound the same, but with varying middle points, include there, their, and they.

The term "variant" applies to words with two different pronunciations and two different meanings. The Greek word is literally a variety of names. Examples based on the same word and various meaning effects are object and object.

A homograph is a word with the same spelling but different pronunciation and different meanings. Other examples are polish and polish, pigeons and pigeons, etc. The term is literally translated as an intermediate word with the same record or description. This word is known as heteronyms and is a term that translates two distinct definitions as an intermediate point.

Popular grammar typesetting close to pandas eating, stinging, and wiping leaves, and the title of a joke that strikes the tongue, are examples of Xerox of homophones that can function as verbs and nouns with a wide variety of effects. In this case, the same word and various meanings of "shoots" completely change the complexion of the sentence.

The comical influences of costumes, puns, and malapropisms have been known for centuries. In Shakespeare's "Much Ado About Nothing," the playwright's legend turned "Dogberryism," an unorganized synonym of malapropism, into an upbeat conversation of a patrol woman Dogberry who declared that he had captured two refreshing men during night surveillance. In the same scene, the patrolman planted a timeless one-liner, not capricious, with the famous saying "Compare stinks."

2.3 Exercise

1. What is the relationship between uncertainty and information?
2. Are fuzziness and uncertainty the same? Justify.
3. Give some relevant examples to show words can mean different things to different people.

References

Fry, Edward. *Homonyms, Homographs, and Homophones: What Is the Difference?* Distributed by ERIC Clearinghouse, Washington, DC (1988).

Keefe, Rosanna. *Theories of Vagueness.* Cambridge University Press, Cambridge (2000).

McCorduck, Pamela, and Cli Cfe. *Machines Who Think: A Personal Inquiry into the History and Prospects of Artificial Intelligence.* CRC Press, Boca Raton, FL (2004).

Viechtbauer, Wolfgang. "Accounting for heterogeneity via random-effects models and moderator analyses in meta-analysis." *Zeitschrift für Psychologie/Journal of Psychology* 215, no. 2 (2007): 104–121.

Zadeh, Lotfi Asker. "PRUF: A meaning representation language for natural languages." *International Journal of Man-Machine Studies* 10, no. 4 (1978): 395–460.

3

Membership Functions and Uncertainty

3.1 Introduction

The membership function (MF) is a line that defines how each point in the input space is mapped to a membership value (or a membership hierarchy) between 0 and 1 (Duch, 2005). Input space is sometimes referred to as universal. A nice name for a simple concept.

One of the most surprisingly used examples of fuzzy sets is the tall people set. In this specimen, spiel's universe is all potential heights, from 3 to 9 ft, and the word "tall" corresponds to the line by which man defines tall strata. Given the tall man set's well-defined (sharp) ditch of the classic set, anyone over 6 ft tall can officially say he or she is tall, but such stardom is arguably absurd. Considering that the numbers are on the utopian side, it may make sense to consider the set of all mistakes greater than six, but when you want to talk almost close to real people, you deny that one is short and the other is the taller. Turning over is irrational. The height is the width of the hair.

But if the kind of stardom that whilom shows doesn't work, what's the right way to pinpoint tall people? Like the plot of the weekend, the pursuit of icons shows a smooth variety of lines, from not tall to tall. The output axis is a number known as a membership value between 0 and 1. Lines are known as membership functions and are often specified as μ. This line defines the transition from not high to high. The two are somewhat tall, but one is much smaller than the other.

The subjective interpretation and objective units are built right into the fuzzy set. If I say "she's tall," the membership feature "tall" should be worth a shot, no matter if I'm 6 years old or an adult. Likewise, units are included in the curve. It makes no sense to say, "Is her height in inches or meters?"

The only condition that the membership function must actually meet is that it must be different between 0 and 1. The function itself can be a misdirected line that can be pinpointed as a function that works for us in terms of simplicity, convenience, and convenience, speed and efficiency.

A classical set might be expressed as

$$A = \{x | x > 6\}$$

A fuzzy set is an extension of a classical set. If X is the universal set and its elements are denoted by x, then a fuzzy set A in X is specified as a set of ordered pairs:

$$A = \{x, \mu A(x) | x \, \varepsilon \, X\}$$

$\mu A(x)$ is tabbed by the membership function (or MF) of x in A. The membership function maps each element of X to a membership value between 0 and 1.

Fuzzy Logic Toolbox includes 11 types of seat membership functions. These 11 functions consist of several important functions such as partial linear function, Gaussian distribution function, sigmoid curve, and quadratic and cubic polynomial curves (Fulcher, 2008). By convention, all membership functions are at the end of their names.

The simplest membership feature works using straight lines. The simplest of these is the trigonometric membership function, and the function name is trimf. It is nothing more than a confusion of the three points that make up a triangle. The trapezoidal membership function trapmf has an appetizing top and is actually a truncated triangular curve. These straight-line membership features have a sound box of simplicity.

Two membership functions are built into the Gaussian distribution curve: a simple Gaussian line and a double-sided composite of two different Gaussian curves. The two functions are gaussmf and gauss2mf.

It is a polynomial-based curve that can be used for multiple membership functions in the toolbox. The three related membership functions are the Z, S, and Pi curves, all named for their shape. The zmf function is an asymmetric polynomial line that is not closed to the left, smf is a mirror image function that opens to the right, and pimf is zero at both extremes rising from the middle.

There are so many different choices that you can separate when choosing your favorite membership features. Also, Fuzzy Logic Toolbox allows you to create your own membership function if you find this list to be too restrictive (Azar, 2010). On the other hand, if this list looks confusing, remember that just one or two types of membership functions, such as triangular and trapezoidal functions, can fit well. There are plenty of options for those looking to explore the possibilities, but a perfectly good fuzzy inference system doesn't require exotic membership features. Member features summary:

- Fuzzy set describes an ambiguous concept (fast runner, hot weather, weekend).
- Fuzzy sets allow the possibility of being a partial member. (Friday is a kind of weekend, and the weather is rather hot.)

- The tier to which the object belongs to the fuzzy set is displayed as a member value between 0 and 1. (Friday is the weekend for tier 0.8.)
- A membership function associated with a given fuzzy set maps input values to membership values.

3.2 Type 1 Membership Function

The type 1 membership function for fuzzy set A of universal set X is specified as $\mu A: X \rightarrow [0,1]$, where each element of X maps to a value between 0 and 1. This value, the tab membership value, or the hierarchy of members quantifies the class of element members of X for fuzzy set A.

The membership function allows you to immerse yourself in the graphic representation of the fuzzy set (Colliot et al., 2006). The x rotation represents the world of discourse, and the y rotation represents the member class of the interval [0,1].

A simple function is used to build the membership function. Given that you are defining a fuzzy concept, using more and more features does not improve accuracy.

3.2.1 The Concept of a Type 2 Fuzzy Set

Membership values may include uncertainty (Douglas et al., 1998). If the value of the membership function is provided by a fuzzy set, it is a type 2 fuzzy set. This concept can be extended to a type n fuzzy set.

3.2.2 Definition of Type 2 Fuzzy Sets and Related Concepts

Type 2 fuzzy sets allow us to incorporate uncertainties close to the membership function into fuzzy set theory, which is a way to write a whilom critique of type 1 fuzzy sets head-on. And if there is no uncertainty, the type 2 fuzzy set is reduced to a type 1 fuzzy set, which is consistent with a decrease in the probability of determinism when the unpredictability disappears. Type 2 purge sets and systems generalize to standard type 1 purge sets and systems to deal with increasingly uncertainty. From the very days of the fuzzy set, there has been little criticism for the fact that the membership function of a type 1 fuzzy set has no uncertainty associated with it. It seems absurd because it has the following meaning: a lot of uncertainty.

A tilde symbol is placed over the symbol of the fuzzy set to symbolically distinguish the type 1 fuzzy set from the type 2 fuzzy set. Thus, A stands for a type 1 fuzzy set, while Ã stands for a comparable type 2 fuzzy set. When the latter is complete, the resulting type 2 purge set is tapped into an unqualified type 2 purge set (to distinguish it from the special spacing type 2 purge set).

Professor Zadeh didn't stop with a type 2 fuzzy set, considering he generalized all of this to a type n fuzzy set in his 1976 paper. We are currently focusing only on type 2 fuzzy sets, considering that our sales target is the next step in the logical progression from type 1 to type n fuzzy sets. Here $n = 1, 2,...$, but some researchers have to navigate higher.

3.2.3 Type 2 Fuzzy Sets and Examples of FOU

The member function of the unqualified type 2 fuzzy set $\tilde{A}$ is three-dimensional, where the third dimension is the value of the member function at each point in the two-dimensional domain tabbed in the Footprint of Uncertainty (FOU) (Gray, 2011).

The membership function of the unqualified type 2 fuzzy set is three-dimensional (3D). The cross section and others are in the FOU.

For an interval type 2 fuzzy set, the 3D values are the same everywhere, so no new information is included in the three dimensions of the interval type 2 fuzzy set. So, for such a set, the third dimension is ignored and only the FOU is used to describe it. For this reason, spacing type 2 fuzzy sets are sometimes represented as tabs in the first-order uncertainty fuzzy set model, while unqualified type 2 fuzzy sets (with useful three dimensions) are also referred to as secondary.

FOU stands for the ambiguity of the type 1 membership function. It is fully described as two boundary functions: LMF (lower membership function) and UMF (upper membership function), both of which are type 1 fuzzy sets! As a result, type 1 fuzzy set math can be used to type and work with interval type 2 fuzzy sets. In this way, engineers and scientists who once knew type 1 fuzzy sets would not have to spend a lot of time learning almost unqualified type 2 fuzzy set math to understand and use interval type 2 fuzzy sets.

Work on type 2 fuzzy sets withered during the early 1980s and mid-1990s, despite the fact that a handful of products were almost published. Since people were still contemplating what to do with a type 1 fuzzy set, the plane was proposed by Zadeh in 1976 as a type 2 fuzzy set, but it wasn't time to give up what the researchers did with the type 1 fuzzy set. This came back in the late 1990s as a result of research by Professor Jerry Mendel and his students on type 2 fuzzy sets and systems. Since then, more and more researchers have been creating products close to type 2 fuzzy sets and systems almost worldwide.

3.2.4 Upper and Lower Membership Functions

I assume that the level of information is not irreversible in specifying the member functions exactly. For example, we can only know the upper and lower premises of the membership class for each element of the universe for a fuzzy set. These fuzzy sets are described as interval value membership functions. For a particular factor x=z, membership is fuzzy set A. That is, µA (z) is expressed as the membership interval [α_1, α_2]. A fuzzy set with an interval value can be generalized further by allowing that interval to fuzzy wither. Then, each membership interval becomes a regular fuzzy set.

3.2.5 A Type 1 Purge Set Represented by a Type 2 Fuzzy Set

A type 1 fuzzy set can be interpreted as a type 2 fuzzy set with all of the second ranks being single (i.e. all flags being 1). In fact, a type 1 fuzzy set is an instance of a type 2 fuzzy set. This is a clear version of the type 2 fuzzy set. Given this case, it makes sense to consider using the type 1 definitions for joins, intersections, and complements as a starting point and generalizing them to a type 2 fuzzy set, which is a type 1 fuzzy set.

3.2.6 0 and 1 Membership of Type 2 Fuzzy Set

The fuzzy set's membership function is a generalization of the classic set's indicator function (Chatzis et al., 2000). In fuzzy logic, it represents the layer of truth as an extension of valuation. Ambiguous truth is conceptually distinguished when considering that it is a member of an ambiguously specified set rather than the likelihood of some event or condition, but the degree of truth often falls with the probability. The membership feature was introduced by Zadeh in his first paper on fuzzy sets (1965). Zadeh suggested in fuzzy set theory to use a membership function (with range tent (0,1)) that works in the domain of all possible values.

3.3 Back to the Language Label

A crisp set of input data is collected and transformed into a fuzzy set using fuzzy language variables, fuzzy language terms, and membership functions (Hong et al., 1996). This step is called fuzzification. The measured (sharp) input is first converted from a fuzzy set to a fuzzy set, taking into account that it is a fuzzy set and not a number that activates the rule described as a non-numeric fuzzy set. Three types of purge fire are available for the interval type 2 FLS. If the measurements are

- perfect, then the noise is modeled as a crisp set.
- noisy, then the noise is fixed and modeled as a type 1 purge set, and
- noisy, then the noise is abnormal and modeled as a gap type 2 purge set (the latter type of purge cannot be cleaned in a type 1 FLS)

3.4 Exercise

1. What is the Footprint of Uncertainty?
2. Distinguish between type 1 and type 2 fuzzy sets.
3. How can a type 2 membership function be considered as an improved version of a type 1 membership function?
4. Write a short note on membership functions.

References

Azar, Ahmad Taher. "Adaptive neuro-fuzzy systems." *Fuzzy Systems* 42, no. 11 (2010): 85–110.

Chatzis, Vassilios, and Ioannis Pitas. "A generalized fuzzy mathematical morphology and its application in robust 2-D and 3-D object representation." *IEEE Transactions on Image Processing* 9, no. 10 (2000): 1798–1810.

Colliot, Olivier, Oscar Camara, and Isabelle Bloch. "Integration of fuzzy spatial relations in deformable models: Application to brain MRI segmentation." *Pattern Recognition* 39, no. 8 (2006): 1401–1414.

Douglas, Andrew P., Arthur M. Breipohl, Fred N. Lee, and Rambabu Adapa. "Risk due to load forecast uncertainty in short term power system planning." *IEEE Transactions on Power Systems* 13, no. 4 (1998): 1493–1499.

Duch, Wlodzislaw. "Uncertainty of data, fuzzy membership functions, and multi-layer perceptrons." *IEEE Transactions on Neural Networks* 16, no. 1 (2005): 10–23.

Fulcher, John, and Lakhmi C. Jain (Eds). *Computational Intelligence: A Compendium*, Vol. 21. Springer, Warsaw, Poland (2008).

Gray, Alexander Westley. "Enhancement of set-based design practices via introduction of uncertainty through use of interval type-2 modeling and general type-2 fuzzy logic agent based methods." PhD dissertation, University of Michigan (2011).

Hong, Tzung-Pei, and Chai-Ying Lee. "Induction of fuzzy rules and membership functions from training examples." *Fuzzy Sets and Systems* 84, no. 1 (1996): 33–47.

4

Case Studies

4.1 Introduction

Case studies refer to in-depth and detailed studies of individuals or small groups of individuals (Crowe et al., 2011). These studies are generally qualitative in nature and lead to descriptive clarifications of policies or experiences. Sample research studies are not used to determine rationalization and effectiveness, nor are they used to discover or predict generalizable truths. Rather, the voice of sample study research lies in the exploration and explanation of the phenomenon. The main characteristics of sample research studies are that they focus narrowly, provide high-level details, and allow for in-depth understanding by combining objective and subjective data.

Quantitative research is willing to ask questions about who, what, where, how much, and how many (Mitchell, 1946). On the other hand, sample studies are used for pun questions about how or why. It is phenomenal for researchers to collect in-depth data in a natural environment where there is little or no lease for events and real-world situations exist. Often the goal of a sample study is to provide information that can be studied in the modification of proposals for future research. Sample studies are phenomenal in social science research and educational settings. For example, a sample study could be used to study psychological issues, such as the trivialities of a child raised by a single parent with a hearing impairment, or the impact on a child who is isolated, lengthy, and neglected by age 12. In addition, sample studies can be used in an educational setting to explore the smallest parts of writing skills in a small group of high school freshmen taking creative writing classes.

There are several types of sample research methods. The method you choose depends on the nature of the content of the question and the researcher's goals. Pursuit is a list of different types of sample studies:

- Illustrative: This type of method is used to "describe" or describe an event or situation in a way that allows people to become more and more familiar with the subject matter in question and, perhaps, wither using terms related to the subject.

- Exploratory: This method is a compressed sample study, and the purpose is to collect basic initial data that can be used to identify specific questions for large-scale studies. This study is not designed to generate detailed data from which conclusions can be drawn. It's essentially simply exploratory.

- Accumulation: The accumulative method is designed to gather information about multiple events/situations and increase the volume in a way that allows for greater generalization. There is a battleship that can save your time and money by avoiding new and repetitive research.

- Key Cases: This study is used to investigate situations of unique interest or challenge common or generalized beliefs. Such research does not create a new generalization. Rather, you can investigate multiple situations or events to raise questions or challenge previously asserted claims.

Once the question is identified and an important type of sample study method is selected, the researcher must unravel the task to design the sample study approach. To get a complete and detailed picture of a participant or a small group, researchers can use a variety of approaches and methods to collect data. These methods may include interviews, field studies, analysis of protocols or transcripts, observing participants without comments, reviewing documents and archived records, and exploring artifacts. Researchers can use either separate (single-method approach) or combined methods (multimodal approach) to collect data (Olsen, 2004).

If the researcher is not swayed by the data mishmash method and what types of data to use and record in the study, data analysis strategies must be determined. Sample research researchers typically interpret data as a whole or through coding procedures. Holistic flow reviews all data as a whole and attempts to draw conclusions based on the data. This makes the content of the question studied to become increasingly clear in nature, and the data tends to provide an overview. Sometimes breaking the data into smaller pieces can become more and more useful. This typically involves searching for data to identify and classify specific behaviors or traits. These categories can prescribe numerical laws by which data can be analyzed using statistical and quantitative methods.

Regardless of the type of sample study, the data mishmash method, or the data weaving method, all sample studies have advantages and disadvantages. The tracking list outlines the potential benefits and limitations associated with using the sample study method.

Advantages:

- Case studies become more flexible than many other types of research and allow researchers to discover and explore as their research progresses.

- Case studies emphasize in-depth content. Researchers are wise to delve deeper and use a variety of data sources to get a well-organized picture.
- Data is calm in its natural environment and context.
- Case studies often lead to a grand universe of new hypotheses that can be tested later.
- Case studies shed new light on established theories that often lead to further exploration.
- Researchers are worldly at studying and solving riddles of situations, events, and actions that can be created in a laboratory environment.

Disadvantages:

- The uniqueness of data is generally the way replication is not worldly wise.
- Case studies have some degree of subjectivity, and researcher bias can be an issue.
- Self-mastery of research on a large scale is impossible due to the deep nature of the data.
- There are very few problems with reliability.

4.2 Time Series Prediction

Forecasting is a method or technique for estimating the future aspects of a trader or operation. It's a way to transform past data or wisdom into future estimates. It is a tool to help executives cope with the uncertainty of the future. Prediction is important for short- and long-term decisions (Abdullah, 2003). Companies can use forecasting in several areas, including technical forecasting, economic forecasting, and demand forecasting. There are two wholesale categories of forecasting techniques: quantitative methods (objective approaches) and qualitative methods (subjective approaches). Quantitative forecasting methods are based on the texture of past data and can use past patterns in the data to predict future data points. Qualitative prediction techniques use the judgment of experts in a specific field to generate predictions (Caniato et al., 2011). They are based on educated guesses or opinions from experts in the field. There are two types of quantitative methods: time series methods and explanatory methods.

The time series method makes predictions based only on historical patterns in the data. The time series method uses time as a self-holding variable to generate demand. In a time series, measurements are taken over a continuous point or over a continuous period. Measurements can be taken hourly, daily, weekly, monthly, yearly, or at other regular (or irregular) intervals. The first step in using a time series tide is to collect historical data. Historical data represent expected conditions in the future. Time series models are an irreversible forecasting tool when demand has shown patterns of outcomes that are expected to repeat in the future. However, looking at data from the past years, you can see that sales of new homes gradually increase over time. In this sample, there is an increasing trend of new home sales. The time series model is divided into four components: a trend component, a periodic component, a seasonal component, and an irregular component. Trends are an important characteristic of time series models (Enders et al., 1992). A time series may be a strike trend, but there may be a trend line where the data points are not swirling or rising. A sequence of repeating points in which a trend line lasting longer than a year is repeated and has not risen is considered to constitute the periodic component of the time series. In other words, these observations in the time series deviate from the trend due to fluctuations.

It provides a good example of a time series that displays a periodic behavior. Components of the time series that capture the variability of the data due to seasonal fluctuations are tabbed in the seasonal component. The seasonal component is similar to the cyclic component in that both refer to constant fluctuations in the time series (Beveridge et al., 1981). The seasonal component captures patterns of regular fluctuations in the time series throughout the year. Seasonal products are more important examples of seasonal ingredients. Random fluctuations in a time series are represented by irregular components. Irregular components of a time series cannot be predicted in advance. Random fluctuations in the time series are caused by short-term and unexpected non-repetitive factors that confuse the time series.

The smoothing method (stable series) is when the time series does not show any significant influence of trend, periodic or seasonal components. In this case, the goal is to smooth out the irregular components of the time series using an averaging process. The moving-average method is the most widely used smoothing technique. In this method, the prediction is the stereotype of the last "x" observation, where "x" is an appropriate number. Suppose that the predictor wants to generate a three period moving average. In the three-period example, the moving-average method uses the average of the three most recent data observations in the time series as predictions for the next time period. These predicted values for the next period, along with the last two observations from the past time series, create a stereotype that can be used as a prediction for the second period in the future. Three-cycle moving stereotype numbering is described in the tracking table. According to the third moving average, you can predict that in 2008, 2.5 million new homes will be most likely to be sold in the United States (Table 4.1).

In moving averages to generate predictions, predictors can experiment with moving averages of different lengths. The predictors separate the lengths that yield the highest truth about the predictions generated. The weighted

TABLE 4.1

Demonstration of Three-Period Moving Averages

Year	Actual Sale (in million)	Forecast (in million)	Calculation
2003	4		
2004	3		
2005	2		
2006	1.5	3	(4 + 3 + 2)/3
2007	1	2.67	(3 + 2 + 3)/3
2008		2.55	(2 + 3 + 2.67)/3

TABLE 4.2

Demonstration of Weighted Three-Period Moving-Average Method

Year	Actual Sale (in million)	Forecast (in million)	Calculation
2005	2 (0.2)		
2006	1.5 (0.3)		
2007	1 (0.4)		
2008		0.42	(2*0.2 + 1.5*0.3 + 1*0.4)/3

moving-average method is a variation of the moving stereotype approach. In the moving-average method, each observation in the data is weighted equally. The weighted moving-average method stipulates variable weights for observations for the data used in the moving average. Again, suppose that the predictor wants to generate a moving average of three periods. In the weighted moving-average method, the three data points receive various weights that exceed the calculated stereotype. Typically, the most recent observation is weighted at its maximum, and the weight is decremented relative to the previous data value (Table 4.2).

An increasingly pervasive form of the weighted shift stereotype is exponential smoothing. In this method, the weights decrease exponentially as the data ages. Exponential smoothing takes the forecast for the previous period and adjusts it by multiplying the predetermined smoothing constant α (called alpha, starting value is <1) by the difference between the previous forecast and the undoubted demand that occurred during the previous forecast period (Mohammed et al., 2017). Exponential smoothing is mathematically expressed as New forecast = start of previous forecast (actual demand – previous forecast). It can be formulated as $F' = F + \alpha (D - F)$.

Other time series forecasting methods include forecasting using trend forecasting, forecasting using trend and seasonal components, and causal forecasting. The trend forecasting method used the underlying long-term trend of the data time series to predict future values. The trend and seasonal component method uses the seasonal component of the time series to match the trend component. The causal method uses a causal relationship between a variable whose future value is predicted and other related variables or factors. A well-known causal method is tap regression, a statistical technique used to develop mathematical models that show how a set of variables are related. You can use this mathematical relationship to generate predictions. There are more and more different time series technologies, such as the ARIMA and Box–Jenkins models. It is a more robust statistical routine that can cope with trending and seasonal data.

Time series models are used in finance to predict the performance or interest rate forecasts of stocks used for weather forecasting. The time series method is probably the simplest distribution method, and it can be very accurate, especially in the short term. Various computer software programs are misogynists to find solutions using time series methods.

4.2.1 Extracting Rules from Data

In theory, neural networks and fuzzy systems are identical in that they are convertible, but in practice, each has its own advantages and disadvantages. In the case of neural networks, knowledge is not automatically invented by the back-propagation algorithm, but the learning process is relatively slow and the texture of the trained network is difficult. Meanwhile, since the input space of the fuzzy system must be divided into fuzzy regions, it is very difficult to wield the fuzzy system to a problem with a large number of input variables. Moreover, difficulties arise in conquering knowledge through expert interviews. However, in the case of fuzzy systems, once knowledge is gained, how the system works is relatively understood. Several methods have been developed to extract fuzzy rules from numerical data to fill the knowledge conquest gap between the two technologies. One Tideway uses a neural network to iterate over fuzzy rules. For example, neural networks are extended to automatically periscope fuzzy rules on numeric data. The main limitation of this method is that you have to specify the number of divisions for each input variable in advance. It also requires splitting of the output variable, but this is not a downside because it can be an unshakable hand depending on the property accuracy. Flipside Tideway extracts fuzzy rules directly from numerical data. For example, fuzzy rules are derived by dividing the input space into fuzzy regions and the output space into regions, and by determining the fuzzy region containing each numerical input data and the output region containing each output data. Fuzzy rules with variable fuzzy regions are extracted for naming problems. This tidal flat solves the problem of the purge system described above. The entry area for each admission is displayed as a series of hyperboxes; overlap of hyperboxes between each other for the same admission is allowed, but overlap between different classes is not. If the data is in the hyperbox belonging to the class, it is judged as that class. The learning algorithm dynamically expands, splits, and contracts hyperboxes when multiple classes overlap.

4.2.2 Classic Time Series Forecasting Method

Machine learning methods can be used for nomenclature and predictions for time series problems. Before delving into machine learning methods for time series, it's a good idea to check if there is an old classic linear time series prediction method. Classical time series prediction methods can focus on linear relationships, but they are sophisticated and perform well on a wide range of problems. The data are well prepared, and the methods are well structured.

The 11 different classic time series prediction methods are as follows:

1. Autoregression (AR)
2. Moving average (MA)
3. Autoregressive moving average (ARMA)
4. Autoregressive integrated moving average (ARIMA)
5. Seasonal autoregressive integrated moving average (SARIMA)
6. Seasonal autoregressive integrated moving average (SARIMAX) using exogenous regression variables
7. Vector autoregression (VAR)
8. Vector autoregressive moving average (VARMA)
9. Vector autoregressive moving average (VARMAX) using exogenous regression variables
10. Simple exponential smoothing (SES)
11. Holt–Winters exponential smoothing (HWES)

4.2.2.1 Autoregression (AR)

The autoregressive (AR) method models the next step in a sequence as a linear function of the observations in the previous time step. The notation for the model involves specifying the order of the model p as a parameter to the AR function: AR (p). For example, AR(1) is a first-order autoregressive model. This method is suitable for univariate time series without trend and seasonal components.

4.2.2.2 Moving Average (MA)

The moving-average (MA) method models the next step in a sequence as a linear function of the residual error of the midpoint process at the previous time step. The moving stereotype model varies from moving stereotypes in a time series. The notation for the model involves specifying the order of the model q as a parameter to the MA function: MA (q). For example, MA (1) is a first-order moving stereotype model. This method is suitable for univariate time series without trend and seasonal components.

You can use ARMA admissions to create an MA model and establish a zero-order AR model. You need to specify the order of the MA models in the order argument.

4.2.2.3 Autoregressive Movement Average Type (ARMA)

The autoregressive moving-average (ARMA) method models the next step in a sequence as a linear function of observations and regression errors in the previous time step, combining autoregression (AR) and moving-average (MA) models. It involves specifying the order of the AR(p) and MA(q) models as parameters to the ARMA function: ARMA (p, q). AR or MA models can be developed using ARIMA models. This method is suitable for univariate time series without trend and seasonal components.

4.2.2.4 Autoregressive Integrated Moving Average (ARIMA)

The autoregressive integrated moving-average (ARIMA) method models the next step in a sequence as a linear function of the difference observation and residual error in the previous time step. We combine the autoregressive (AR) and moving stereotype (MA) models and the difference preprocessing steps of the sequence to make the sequence a fixed, tap-integrated (I). Model notation involves the ordering of AR. It takes the models (p), I(d), and MA(q) as parameters to the ARIMA function: ARIMA (p, d, q). You can also use ARIMA models to develop AR, MA, and ARMA models. This method is suitable for univariate time series with trends and no seasonal components.

4.2.2.5 Seasonal Autoregressive Integrated Moving Average (SARIMA)

The seasonal autoregressive integrated moving stereotype (SARIMA) method models the next step in a sequence as a linear function of difference observations, errors, seasonal difference observations, and seasonal error from the previous time steps. Model notation involves specifying the order of the AR(p), I(d), and MA(q) models as parameters for ARIMA functions and AR(P), I(D), MA(Q), and m. In SARIMA (p, d, q) (P, D, Q) m , "m" is the number of time steps in each season (seasonal period). AR, MA, ARMA, and ARIMA models can be developed using SARIMA models, which are suitable for univariate time series with trend and/or seasonal components.

4.2.2.6 Seasonal Autoregressive Integrated Moving Average (SARIMAX) Using Exogenous Regression Variables

Seasonal autoregressive integrated moving-average with exogenous regressors (SARIMAX) is an extension of the SARIMA model that also includes modeling of exogenous variables. Exogenous variables are also tab-delimited covariates and can be thought of as parallel input sequences with observations at the same time step as the original series. The primary series can be said to be endogenous data that are not related to the exogenous sequence(s). Observations of exogenous variables are included in the model directly at each time step and are not modeled in the same way as primary endogenous sequences (e.g., AR and MA processes). It can also be modeled using the SARIMAX method. Included models with exogenous variables are ARX, MAX, ARMAX and ARIMAX. This method is suitable for univariate time series with trend and/or seasonal components and exogenous variables.

4.2.2.7 Vector Autoregression (VAR)

The vector autoregression (VAR) method uses an AR model to model the next step in each time series. It is a generalization of AR to multiple parallel time series. Multivariate time series: Model notation involves specifying the order of the AR(p) model as a parameter to the VAR function, e.g. VAR(p). This method is suitable for multivariate time series without trend and seasonal components.

4.2.2.8 Vector Autoregressive Moving Average (VARMA)

The vector autoregressive moving-average (VARMA) method uses an ARMA model to model the next step in each time series. It is a generalization of ARMA to multiple parallel time series. Multivariate time series: Model notation involves specifying the order of the AR (p) and MA (q) models as parameters to the VARMA function – VARMA (p, q). You can also use VARMA models to develop VAR and VMA models. This method is suitable for multivariate time series without trend and seasonal components.

4.2.2.9 Vector Autoregression Moving-Average with Exogenous Regressors (VARMAX)

The vector autoregression moving-average with exogenous regressors (VARMAX) is an extension of the VARMA model that also includes the modeling of exogenous variables. It is a multivariate version of the ARMAX method. Exogenous variables are also tabbed covariates and can be thought of as parallel input sequences that have observations at the same time steps as the original series. The primary series(es) are referred to as endogenous data to distinguish from the exogenous sequence(s). The observations for exogenous variables are included in the model directly at each time step and are not modeled in the same way as the primary endogenous sequence (e.g. AR and MA processes). The VARMAX method can also be used to model the subsumed models with exogenous variables, such as VARX and VMAX. The method is suitable for multivariate time series without trend and seasonal components and exogenous variables.

4.2.2.10 Simple Exponential Smoothing (SES)

The simple exponential smoothing (SES) method models the next time step as an exponentially weighted linear function of observations at prior time steps. The method is suitable for univariate time series without trend and seasonal components.

4.2.2.11 Holt–Winters Exponential Smoothing (HWES)

The Holt–Winters Exponential Smoothing method comprising of the forecast equation and three smoothing equations models the next time step as an exponentially weighted linear function of observations at prior time steps, taking trends and seasonality into account. The method is suitable for univariate time series with trend and/or seasonal components.

4.3 Knowledge Mining Using Surveys

Data mining is the process of discovering patterns in large datasets involving methods at the intersection of machine learning, statistics, and database systems. Data mining is an interdisciplinary subfield of computer science and statistics with an overall goal to periscope information (with intelligent methods) from a dataset and transform the information into a comprehensible structure for remoter use. Data mining is the wringer step of the "knowledge discovery in databases" process, or KDD. Aside from the raw wringer step, it also involves database and data management aspects, data pre-processing, model and inference considerations, interestingness metrics, complexity considerations, post-processing of discovered structures, visualization, and online updating. The difference between data wringer and data mining is that data wringer is used to test models and hypotheses on the dataset, e.g., analyzing the effectiveness of a marketing campaign, regardless of the value of data, whereas data mining uses machine learning and statistical models to uncover underhand or subconscious patterns in a large volume of data.

The term "data mining" is in fact a misnomer, considering that the goal is the extraction of patterns and knowledge from large amounts of data, not the extraction (mining) of data itself. It is also a buzzword and is wontedly unromantic to any form of large-scale data or information processing (collection, extraction, warehousing, analysis, and statistics) as well as any using of computer visualization support system, including strained intelligence (e.g., machine learning) and merchantry intelligence. Often the increasingly unstipulated terms (large-scale) data wringer and analytics – or, when referring to very methods, strained intelligence and machine learning – are increasingly appropriate.

The very data mining task is the semi-automatic or will-less wringer of large quantities of data to periscope previously unknown, interesting patterns such as groups of data records (cluster analysis), unusual records (anomaly detection), and dependencies (association rule mining, sequential pattern mining). This usually involves using database techniques such as spatial indices. These patterns can then be seen as a kind of summary of the input data, and may be used in remoter wringer or, for example, in machine learning and predictive analytics. For example, the data mining step might identify multiple groups in the data, which can then be used to obtain increasingly well-judged prediction results by a visualization support system. Neither the data collection, data preparation, nor result interpretation and reporting is part of the data mining step, but do vest to the overall KDD process as spare steps.

The related terms "data dredging," "data fishing," and "data snooping" refer to the use of data mining methods to sample parts of a larger population dataset that are (or may be) too small for reliable statistical inferences to be made well-nigh the validity of any patterns discovered. These methods can, however, be used in creating new hypotheses to test versus the larger data populations.

In view of the tremendous production of computer data worldwide, there is a strong need for new powerful tools that can automatically generate useful knowledge from a variety of data, and present it in human-oriented forms. In efforts to satisfy this need, researchers have been exploring ideas and methods ripened in machine learning, statistical data analysis, data mining, text mining, data visualization, pattern recognition, etc. A multi-strategy methodology for an emerging research direction, tabbed knowledge mining, is a tool by which we midpoint the derivation of high-level concepts and descriptions from data through symbolic reasoning involving both data and relevant preliminaries knowledge. The constructive use of preliminaries as well as previously created knowledge in reasoning well-nigh new data makes it possible for the knowledge mining system to derive useful new knowledge not only from large amounts of data, but also from limited and weakly relevant data.

4.3.1 Knowledge Mining Methodology

Step 1 concerns the proper preparation of data for modeling, handling of missing values, and support or minimization of total errors through elimination of outliers and non-real values. Filtering should be applied at this stage, that is, choosing the type and range of data to be analyzed. For example, it consists of selecting a variety of specific products, or often narrowing down data across millions of data records. Data transformation should be non-romantic (mostly consist of normalization or normalization of the data) if necessary for the use of a specific DM method. For example, the use of MIN-MAX regularization at the model tower step is required for samples using neural networks where it is recommended, reducing the risk of redundant effects of data size on model results.

In the second stage, software is mainly used. Certain data mining methods are implemented to perform exploration tasks. At this point, the dataset prepared in step 1 is used. The speed of working with the database and tower of the model depends not only on the complexity of the problem itself, but also on the type, value, and weft of the data (qualitative or quantitative data, models obtained with or without a teacher, the number of data records, and the dependency between input and output variables). This stage of use of the data mining method is most surprisingly automatic, while the realization time is related to the complexity of the mentioned problem, but to the performance of the computing equipment itself. While the data mining solver doesn't require a powerful graphics card or large, inflexible disks (more and more often data distribution and grid computing methods are used), it's important to have a good CPU and a large RAM value. However, a powerful computer device with properly configured graphic sebum can significantly push computing.

Step 3 is about the interpretation of the results obtained in step 2. It is important to participate in the trials, gain expertise in Statistical Methods and Data Mining. Step 2 may seem relatively simple using the capabilities of a modern computer, but other steps require specialized knowledge. Enterprises realize the ripple effect of data exploration projects that require the cooperation of experts from multiple branches and visitor departments. In the literature, various data mining methodologies are proposed in the form of scenarios for collecting and preparing data for telemetry and distributing the results for the implementation of specific solutions.

4.4 Exercise

1. What do you mean by time series?
2. Describe various components of time series.
3. What is the difference between data mining and knowledge mining?
4. Describe various stages of knowledge mining.
5. Justify the use of Python in time series analysis.

References

Abdullah, Ramli. "Short term and long term projection of Malaysian palm oil production." *Oil Palm Industry Economic Journal* 3, no. 1 (2003): 32–36.

Beveridge, Stephen, and Charles R. Nelson. "A new approach to decomposition of economic time series into permanent and transitory components with particular attention to measurement of the 'business cycle'." *Journal of Monetary Economics* 7, no. 2 (1981): 151–174.

Caniato, Federico, Matteo Kalchschmidt, and Stefano Ronchi. "Integrating quantitative and qualitative forecasting approaches: Organizational learning in an action research case." *Journal of the Operational Research Society* 62, no. 3 (2011): 413–424.

Crowe, Sarah, Kathrin Cresswell, Ann Robertson, Guro Huby, Anthony Avery, and Aziz Sheikh. "The case study approach." *BMC Medical Research Methodology* 11, no. 1 (2011): 100.

Enders, Walter, Gerald F. Parise, and Todd Sandler. "A time-series analysis of transnational terrorism: Trends and cycles." *Defence and Peace Economics* 3, no. 4 (1992): 305–320.

Mitchell, Wesley C. "Empirical research and the development of economic science." In Arthur F. Burns (Ed.) *Economic Research and the Development of Economic Science and Public Policy*, pp. 1–20. NBER, Cambridge, MA (1946).

Mohammed, Jameel, Sanjay Bahadoorsingh, Neil Ramsamooj, and Chandrabhan Sharma. "Performance of exponential smoothing, a neural network and a hybrid algorithm to the short term load forecasting of batch and continuous loads." In *2017 IEEE Manchester PowerTech*, pp. 1–6. IEEE, Manchester (2017).

Olsen, Wendy. "Triangulation in social research: Qualitative and quantitative methods can really be mixed." *Developments in Sociology* 20 (2004): 103–118.

5

Singleton Type 1 Fuzzy Logic Systems: No Uncertainties

5.1 Introduction

There is uncertainty in every real application, and when faced with high levels of uncertainty, its performance may not be true. So it's important that these usage systems are worth dealing with these uncertainties. Fuzzy set (FS) and fuzzy logic system (FLS) are existing concepts and techniques that have been chosen as the methodology of tower systems that can unleash new performance in uncertainty and inaccuracies. There are many sources of uncertainty faced by FLS, such as the presence of noise in training data, well-represented inputs and outputs of noise, and linguistic uncertainty related to the linguistic terminology of rule based protectors (Hagras, 2007).

The non-singleton fuzzy logic system (NSFLS) was introduced to model the uncertainty of the input signal as an extension of the singleton fuzzy logic system (SFLS), which is not an issue when the input data is corrupted due to measurement noise. In NSFLS, the inputs are modeled as FSs and are no longer well-performing values. NSFLS has been used successfully in a variety of applications, and new advances in the development of new types of NSFLS have shown excellent results (Aladi et al., 2016).

5.2 Rules

Fuzzy rules are the confusion of verbal statements that describe how a fuzzy inference system (FIS) creates visualizations with regard to classifying inputs or executives as outputs. Fuzzy rules are written unchanged in trace format.

(input1 is membership function1) and/or (input2 is membership function2). Then (output k is the output membership function k).

For example, you could create a rule like "High temperature and high humidity make the room hot."

It should have a membership function that finds out exactly what the upper temperature (input1), upper humidity (input2), and hot room (output1) mean. This process of taking an input like temperature and processing it through a member function to determine the middle point with a "high" temperature is tapped fuzzification. Also, in the fuzzy rule, you need to find exactly the middle point with "and"/"or". This is a tab-delimited fuzzy combination.

5.3 Fuzzy Inference Engine

Sugeno FIS is very similar to Mamdani FIS. The main difference between these is that the output size is not calculated in Sugeno FIS by clipping the output membership function from the rule strength (Lofti, 1995). In fact, Sugeno FIS does not have any output membership features. Instead, the output is a well-performing number calculated by multiplying each input by an unchanging value and then subtracting the result.

In this example, "Rule Strength" is called "Applicability" and the output is called "Action." Also note that there is no output distribution, only "result behavior," which is a mathematical combination of rule strength (applicability) and output (action).

5.4 Fuzzification and Its Effect on Reasoning

Fuzzification is the first step in the fuzzy reasoning process. This includes domain transformations where sharp inputs are converted to fuzzy inputs. Clear inputs are word-by-word inputs measured by sensors and passed to the tenancy system for processing such as temperature, pressure, and rpm. Each well-performed input that the FIU will process has its own group of members. This member functional group exists within a world of discourse that holds all the relevant values that a good input can have. The pursuit shows the structure of membership functions within Spiel's world for well-informed input.

When designing the number of membership functions for the input variable, the label should not be shaken for the membership function in the first place. The number of labels corresponds to the number of areas in which you need to split the universe so that each label describes the area of action.

Telescopic should be provided for each membership function that numerically identifies the range of input values corresponding to the label. The shape of the membership function should be representative of the variable (Thiem, 2014). However, this form is further limited by the available computing resources. Complex shapes require increasingly granular explanatory equations or large lookup tables.

5.4.1 Fuzzifier

Fuzzification is the process of converting sharp inputs into fuzzy values. FLC's data manipulation is based on FS theory, and fuzzification is necessary and desirable at an early stage. Thus, fuzzy fire can be specified as a mapping from the observed input space to the FS label of a given input discourse universe.

The mapping function oversees the associated measurement uncertainty for each input variable. The purpose of the mapping function is to interpret the measurements of the input variables, each expressed as a real number and an increasingly realistic fuzzy approximation of each real number (McBratney et al., 1997).

If f is a non-romantic mapping function to variable x, you can write it as follows:

$$f(x_i):[-k,k] \rightarrow R$$

where R is the set of fuzzy numbers and $f(x_0)$ is the real number chosen by f as the fuzzy characteristic of the measurement $x_i = x_0$. The possible definitions for this fuzzy number for all $x_i \in [-a, a]$ are shown in the icon below. If desired, you can use a different shaped mapping function for the fuzzy number $f(x_0)$. For each measure $x_i = x_0$, an FS $f(x_0)$ enters the inference mechanism.

5.5 Defuzzification

Defuzzification is the process of performing output renormalization that converts a set of controller output values to single point values and maps the point values of the controller output to the physical domain (Eatherley et al., 1994). Conceptually, the job of the defuzzifier is to assign a weighted point to W with the FS C' obtained as a result of the inference engine. Then, inverse fuzzification can be specified as a mapping of the set C' to the well-spoken deportation space $z* \in W$ in the fuzzy deportation space $W \subset R$ in the world of speech output.

There are many choices to determine this well-performed controller output $z*$, but when choosing a diffraction scheme, you need to consider the four criteria you seek:

- Validity: The point $z*$ should represent C′ from an intuitive point of view. For example, it could roughly lie in the middle of the support of C′ or have a higher tier of members at C′.
- Calculation simplicity: This criterion is particularly important for fuzzy tenancy given that the fuzzy controller operates in real time.
- Continuity: A small increase in C′ should not have a large increase in $z*$.
- Clarity: This is how the deepening method produces a unique value for $z*$ without change.

5.5.1 Centroid Defuzzifier

This method is known as part of gravity or a region. This technique was matured by Sugeno. This is the most surprisingly used technique. The only downside to this method is that it is difficult to compute for the classified member functions.

5.5.2 Bisecting Defuzzifier

A bisector is a vertical line that divides a region into two subregions of the same region. Sometimes it coincides with the centerline, but not invariably.

5.5.3 Weighted Average Defuzzifier

This method is usually limited to a symmetric output membership function, and the weighted average method is formed by weighting each maximum membership value to each membership function in the output.

5.5.4 Midpoint of Maximum Defuzzifier

In this defuzzy method, only the clever rules with the highest fulfillment tier are considered.

5.5.5 Largest of Maximum Defuzzifier

The largest of the maximum values is the sharpest value tapped with ZLOM, taking the largest of all z bested in $[z_1, z_2]$.

5.5.6 Smallest Maximum Defuzzifier

Select the smallest output with maximum membership function with good value ZSOM. That is, in the smallest of maximum, the smallest of all z given to $[z_1, z_2]$ is selected.

5.6 FLS Design

FLS is a nonlinear system that can infer nonlinear relationships between input and output variables. The nonlinear properties are especially important when there is an underlying physical mechanism. The system can "learn" a nonlinear mapping with a set of input signals that are used in conjunction with an optimization algorithm to determine the system value and what gives the desired response pair: parameter. This is tab-delimited supervised learning, one of the most surprisingly used learning paradigms. Even if the process to be modeled is not static, the system can be updated to reflect the confusing statistics of the process. Unlike traditional probabilistic models used to model these processes, FLS makes no assumptions about the structure of the process and does not call any kind of probability distribution model: nonparametric method (Mendel et al., 1997).

Designing an FLS can be viewed as either approximating a function or (probably) fitting the rupture surface to a higher-dimensional space. Given a set of input–output pairs, the learning task is practically equivalent to determining the system that provides the best fit for the input–output pair in terms of the mole function. Also, the system produced by the learning algorithm must be secular in order to generalize to a specific area of a multidimensional space where no training data are provided. This means that interpolating a given input–output data has to be worldly. It is used worldwide among many approximations/interpolations within the framework of approximation and interpolation theory.

5.6.1 Back-Propagation (the Steepest Descent) Method

Adaptive filtering has been widely used and successful in many areas such as image processing, tenancy, and communication. In classical filtering theory, signal interpretation based on prior knowledge of the desired signal and noise can rely heavily on the model's linearity and fixed mathematical methods. However, for samples with upper nonlinearity, the interpretation of performance may not be acceptable. In such cases, neural fuzzy filtering can provide the greatest solution to the noise filtering problem. Neural networks are well balanced with numerous highly interconnected processing elements (nodes) that do not fluctuate through weights. Looking at the structure and parameter learning of neural networks, you can find many points around the world on methods used for adaptive signal processing. The back-propagation algorithm used for training neural networks is the generalized Widrow's least-medium-squares (LMS) algorithm and can be contrasted with the LMS algorithm commonly used for adaptive filtering.

Many kinds of nonlinear filters designed using neural networks have been proposed. One of them is the neural filter, which makes learning algorithms appear increasingly more efficient than Lin's adaptive stack filtering algorithm. The input of this neural filter is based on the threshold decomposition and neural network, and it is divided into an inflexible neural filter (an energetic function is a unit step) and a soft neural filter (an animate function is a sigmoid function). Another kind of neural filter is a recursive filter obtained by learning a recurrent multilayer perceptron (RMLP). Other applications of neural networks in adaptive filtering include nonlinear constant equalizers and noisy speech recognition. Here, the neural network is used to map the noisy input function to the wipe output function for recognition. The problem faced by the neural filter's diamond is that it is difficult to determine the structure and size of the network as the inner layer of the neural network is invariably opaque to the user.

Several approaches have been proposed to enable neural networks to learn from numerical data as well as expertise expressed in fuzzy if–then rules. We can overcome the shortcomings of neural filters and present neural fuzzy filters while maintaining their advantages. Neurofuzzy filters use system input and output data to learn system policies, so they do not use mathematical modeling.

After learning the behavior of the system, the neurofuzzy filter automatically generates fuzzy rules and membership functions to solve the main problems of fuzzy logic and significantly reduce diamond trending. Then, the neurofuzzy filter checks the solution (generated rules and membership functions). It also optimizes the number of rules and membership features. Fuzzy logic solutions matured by neurofuzzy filters solve implementation problems related to neural networks. Unlike traditional fuzzy logic, neurofuzzy filters use new deferentialization, rule inference, and source processing algorithms to provide an increasingly reliable and well-determined solution. This new algorithm is mature based on neural network architecture. Finally, the unwilling legislative converter converts the optimized solution (rules and membership functions) into the turnout code of the embedded controller.

5.6.2 SVD-QR Method

An SVD-QR-based assistant is proposed to diamond a fuzzy system directly from some collected input–output data. A fuzzy system with a table of fuzzy rules is assigned to orchestrate the input–output pairs of the identified system. In the rule concatenation of a given fuzzy system, each fuzzy rule table corresponds to a partition in the input space. In order to periscope the most important fuzzy rules in the rule wiring of a given fuzzy system, an unshakable firing intensity matrix is constructed by the membership function of the entire FS. As with the firing intensity matrix, the number of important fuzzy rules is not shaken by a singular value decomposition (SVD), and the most important fuzzy rules are chosen by the SVD-QR-based method.

As a result, the reconstructed fuzzy rule wiring with a well-balance of important fuzzy rules is not shaken by the firing intensity matrix. In addition, the recursive least squares method is unromantic in determining the resulting portion of the reconstructed fuzzy system, so the fine fuzzy system is not shaken by the proposed method, which is the same as the collected input and output data. Finally, the nonlinear system demonstrates the effectiveness of the proposed tide channel for fuzzy modeling.

5.6.3 Repetitive Diamond Method

Iterative learning control (ILC) improves the performance of a control system (CS) by repeatedly using past wisdom in CS operations. The ILC system iteratively solves the parameter optimization learning problem that minimizes the objective function that specifies CS performance. One way to look at fuzzy tenancy is as a user-friendly early nonlinear assistant dealing with rippled or poorly defined vegetation designed in a heuristic way that incorporates human technology but no universal diamond method.

This poses problems with the structural properties, stability, controllability, parameter sensitivity, and robustness of fuzzy control systems (FCSs). The purpose of combining fuzzy tenancy and ILC is to modernize the FCS's figure of merit taking into account both the benefits of the feedback due to fuzzy tenancy and the feed-forward bounty due to the ILC merged into the same CS structure.

5.7 Sample Study: Time Series Prediction

We are going to use one of the most surprisingly used methods for time series prediction: autoregressive integrated moving average (ARIMA). ARIMA models are denoted in ARIMA (p, d, q) notation. The following three parameters are worthwhile for seasonality, trend, and noise in your data:

```
p = d = q = range(0, 2)
pdq = list(itertools.product(p, d, q))
seasonal_pdq = [(x[0], x[1], x[2], 12) for x in
list(itertools.product(p, d, q))]
print('Examples of parameter combinations for Seasonal
ARIMA...')
```

```
print('SARIMAX: {} x {}'.format(pdq[1], seasonal_pdq[1]))
print('SARIMAX: {} x {}'.format(pdq[1], seasonal_pdq[2]))
print('SARIMAX: {} x {}'.format(pdq[2], seasonal_pdq[3]))
print('SARIMAX: {} x {}'.format(pdq[2], seasonal_pdq[4]))
```

This step is the parameter selection for our furniture sales' ARIMA Time Series Model. Our goal here is to use a "grid search" to find an optimal set of parameters that yield the weightier performance for our model:

```
for param_seasonal in seasonal_pdq:
    try:
        mod = sm.tsa.statespace.SARIMAX(y,
                        order=param,
                        seasonal_order=param_seasonal,
enforce_stationarity=False,
                        enforce_invertibility=False)
results = mod.fit()
print('ARIMA{}x{}12 - AIC:{}'.format(param, param_seasonal,
results.aic))
        except:
            continue
```

The output suggests that SARIMAX(1, 1, 1)×(1, 1, 0, 12) yields the lowest AIC value of 297.78. Therefore, we should consider this to be optimal option.

5.8 Case Study: Data Mining

Data mining (DM) is the extraction of implicit, previously unknown, and potentially useful information from data. It is the use of specific algorithms to extract patterns from the data. When strong patterns are found, they can be generalized to make well-judged predictions about future data.

The pattern found should definitely be valid in the new layered data. We also want the pattern to be new (at least to the system and preferably to the user) and potentially useful. In other words, it provides some benefit to the user or task. Finally, you should be able to understand the pattern immediately without any post-processing.

Knowledge discovery goals are specified according to the purpose of the system. Discovery objectives are divided into predictions, where the system finds patterns to predict the future policy of some entities, and descriptions, where the system finds patterns to display to users in a human understandable form.

The following are the main types of mining:

- Association rule mining (genetic algorithms, CN_2 rules): A rule set is the warmest expression of knowledge and is probably the easiest to understand. This includes searching for rules that are used to describe the conditions under which items occur together. In general, the condition of a rule is a predicate of a specific logic, and whoopee is the associated class. In other words, it predicts whoopee for the input instances that create the true condition. It is often used for market basket or trading data analysis.
- Classification and prediction (CART, CHAID, ID_3, $C_{4.5}$, $C_{5.0}$, $J_{4.8}$): This includes the identification of data features that can be used to generate models to predict similar occurrences in future data. The decision tree categorizes instances by sorting them down the tree, from the root to some leaf nodes, which gives the nomenclature of the instances. Each node in the tree specifies a test for some symbol of the instance, and each workshop descending from that node corresponds to one of a range of possible values for this property.
- Cluster analysis (K-means, neural networks, expectation-maximization): Attempts to squat on a group (cluster) of data items that are very similar to the other objects in the group, but not most similar to the objects in the other groups.

It can be said that in an unqualified way, these algorithms use a limited representation of knowledge. Given the structure of the knowledge to be acquired, there is little possibility of transformation. For example, a rule is just a tentative relationship between a condition and a conclusion. This limits the possibility of acquiring new knowledge, where the main logical relationship is joining, separation or equality, or a self-dominating combination of logical operators.

Each knowledge expression language takes into account the limitations of its definition and limits the space of possible solutions. Any learning algorithm and knowledge representation can be a more weighted algorithm in a subset of non-recurring problems. For this reason, the main purpose of this work is to find new methods for knowledge discovery in databases that allow you to obtain fuzzy predicates in an increasingly flexible way of expressing knowledge using various meta-heuristic algorithms.

The point of view to consider here is a different point of view than what Zadeh described in 1994. For us, the main components are FSs, stochastic inference, neural networks, and meta-heuristics. All in all, these components are complementary, not competitive. It is increasingly known that in many cases, it is worth combining them. The hybridization of the soft computing context favors and reinforces the visit of the original procedure, which can help solve new problems like this. Soft computing will play an increasingly important role in many areas of use. DM methods can be viewed as three main algorithmic components: (1) model representation, (2) model evaluation, and (3) search.

5.9 Exercise

1. Describe how ARIMA method can be utilized for forecasting.
2. What do you midpoint by data mining? Mention a few applications of it.
3. What is defuzzification? Why is it necessary? Describe a few techniques of it.
4. Distinguish between SVDQR and iterative diamond method.
5. Describe the effect of fuzzification on inference.

References

Aladi, Jabran Hussain, Christian Wagner, Amir Pourabdollah, and Jonathan M. Garibaldi. "Contrasting singleton type-1 and interval type-2 non-singleton type-1 fuzzy logic systems." *In 2016 IEEE International Conference on Fuzzy Systems (FUZZ-IEEE)*, Vancouver, Canada (2016), pp. 2043–2050.

Eatherley, Graham J. *Autonomous Vehicle Docking Using Fuzzy Logic.* University of Ottawa, Canada, 1994.

Hagras, Hami. "Type-2 FLCs: A new generation of fuzzy controllers." *IEEE Computational Intelligence Magazine* 2, no. 1 (2007): 30–43.

Lofti, Ahmad. "Learning fuzzy inference systems." PhD dissertation, University of Queensland (1995).

McBratney, Alex B., and Inakwu O.A. Odeh. "Application of fuzzy sets in soil science: Fuzzy logic, fuzzy measurements and fuzzy decisions." *Geoderma* 77, no. 2–4 (1997): 85–113.

Mendel, Jerry M., and George C. Mouzouris. "Designing fuzzy logic systems." *IEEE Transactions on Circuits and Systems II: Analog and Digital Signal Processing* 44, no. 11 (1997): 885–895.

Thiem, Alrik. "Membership function sensitivity of descriptive statistics in fuzzy-set relations." *International Journal of Social Research Methodology* 17, no. 6 (2014): 625–642.

Zadeh, Lotfi. "Fuzzy logic (abstract) issues, contentions and perspectives." In *Proceedings of the 22nd Annual ACM Computer Science Conference on Scaling up: Meeting the Challenge of Complexity in Real-World Computing Applications: Meeting the Challenge of Complexity in Real-World Computing Applications* (1994), p. 407.

6

Centroid of a Type 2 Fuzzy Set: Type Reduction

6.1 Introduction

An unqualified type 2 fuzzy set (T2 FS) is a T2 FS whose secondary membership can have a value between [0,1]. Compared to the spacing T2 (IT2) FS where the secondary membership is all 1, the unqualified T2-FS has more and more diamond-level self-rule and consequently is getting more and more support from T2 FS researchers.

An important numbering for an unqualified T2 FS is centered given that the centroid provides a measure of the uncertainty of such an FS and may have to calculate the centroid during type reduction for an unqualified type 2 fuzzy logic system (T2 FLS). The centerpiece of the unqualified T2 FS matured by Karnik and Mendel is the union of the centers of all built-in T2 FSs. Unfortunately, these exhaustive and unreliable centroid calculations are practically impossible due to the very large number of built-in T2 FSs. Various practical methods have been proposed to calculate or impair the centroid (Greenfield and Chiclana 2013). I proposed to sample multiple embedded T2 FSs randomly and provided some examples showing that the number of randomly selected embedded sets only slightly affects the central midpoint value. Excellent results have been obtained for ten with a small number of randomly selected embedded T2FSs, but it is still not known in theory why the number of randomly selected embedded T2FSs could lead to such good "differentiated" results. This method does not provide a well-formed member function (MF) of the center. Coupland recommended using the x coordinate of the geometric center of A's 3D MF. As his assistant Coupland, he goes directly from A to a well-performing number, not T1 FS. Therefore, it does not provide a measure of the uncertainty of A. Also, when T2 FS decreases to T1 FS (when all uncertainties disappear), this x coordinate does not decrease to the exact center of T1 FS.

It was proposed to calculate the centroid as the centroid of all vertical slices (T1 FS each) (Liu, 2011). This ad hoc method creates an unchanging centroid where the domain of the nonzero centroid value is equal to the domain of the nonzero centroid of A's primary variable, regardless of the remaining geometry of the T2 FS. This doesn't seem to be an accurate measure of A's uncertainty given that the nonzero-valued region of the center of A has to depend on the remaining geometry of the T2 FS. Most recently, Liu showed how to calculate the centroid of A using the α-plane representation theorem (RT) of T2 FS (also tab-delimited horizontal slice RT). He showed that the center of A is the union of the centers of all α-level T2 FS of A. The Karnik–Mendel (KM) algorithm or the enhanced Karnik–Mendel (EKM) algorithm is used to calculate the centroid of each α-level T2 FS. Mendel and Liu have shown that the KM algorithm is monotonous and converges exponentially fast, but requires a few (typically 2–6) iterations beyond the convergence invariably. As a result, finely discretizing α can lead to computational undersongs that can hinder the widespread use of this method due to the iterative nature of the KM (i.e. EKM) algorithm.

The T2 FLS theory and its applications have grown in recent years. One of the major problems associated with the implementation of these systems is type reduction, which computes the generalized centroid of T2 FSs (Liu, 2008). This operation becomes computationally simpler when done for a specific matrix of T2 FS, i.e. an interval type 2 fuzzy set (IT2 FS). Basically, the center of IT2 FS is the gap. So, calculating this centroid can be considered an optimization problem to find the last point that exactly points to the gap. The fast calculation of type reduction for IT2 FS is an attractive problem, and it is a very important problem as the type-reduction procedure for an increasingly unqualified T2 FS uses interval type 2 fuzzy calculations. To date, several iterative approaches have been proposed to reduce the type of computing for IT2 FS.

The KM algorithm is the most used procedure in interval type 2 fuzzy logic systems (IT2 FLSs). It has been proven that the KM algorithm is monotonous and converges very exponentially fast. These properties are repeatedly very desirable. However, the KM procedure converges in multiple iterations and requires significant arithmetic and memory lookup work for an actual IT2 FLS implementation. In a sample of IT2 fuzzy controllers, the calculation forfeiture of type reduction is an important topic. Most of the controllers depend on type reduction and fuzzification steps.

Therefore, it is very convenient to develop strategies to reduce the computational burden and the resources required to implement these two steps. There are also other applications of IT2 FLS that need to reduce the complexity of the hardware and software platform to ensure the fastest possible execution of fuzzy inference. The non-repetitive type reduction of IT2 FS can be achieved through uncertainty boundaries. However, this method is harmful and involves relatively distracting calculations (Torshizi et al., 2015). Other methods of calculating the centroids of IT2 FS have been proposed, such as geometric type reduction, genetically optimized type reducer, spacing analysis-based type reduction, and sampling and decay differentiation. These volitional methods have interesting properties, but are not as popular as the KM algorithm.

6.2 Unspecified Consequences for the Center

The definition of the continuous central function and the study of each are provided by Mendel and Liu (2007). Theoretical work can be justified for a discrete central function. In connection with this fact, Mendel and Liu showed several interesting properties of the continuous center function. It can be summed up that some of these properties have a decrease in central function or there is an appetizing spot to the left of the minimum value, and an increased or an appetizing spot to the right of this value. Also, this function has a global minimum when k=L. The right-center function yr has similar properties because when k=R, this function has a global maximum. Therefore, it increases to the left of the maximum value and decreases to the right.

6.3 Generalization Center for Interval Type 2 Fuzzy Set

Calculating the generalized centroid of the interval type 2 fuzzy set (IT2 FS) is an important step in the operation of the IT2 FLS. Several algorithms have been proposed to calculate this, but the KM algorithm is the most used method in type 2 fuzzy logic applications. The computational properties of the KM algorithm have been improved from, resulting in a new version with the enhanced Karnik–Mendel (EKM) algorithm tab. Another iterative algorithm is proposed for calculating the generalized centroid of IT2 FS. Experimental traces show that this algorithm is faster than the original KM algorithm. However, there is no computational comparison between it and the EKM algorithm. The algorithm is really simple. By running an exhaustive search over the non-shortened central function to find the min and max, you can reduce search time by introducing conditions that stop the algorithm when the minimum or maximum value is reached. The discrete centroid function is not convex, but has well-defined properties, which helps to set the required stopping conditions. This work here introduces a new version of the tabbed algorithm to the iterative algorithm with stop conditions (IASCO) and compares it with the EKM algorithm in terms of computation time and arithmetic precision.

6.4 Interval Type 2 Center of Fuzzy Set

The IT2 FS is the most widely used T2 FS today, given its computational simplicity. When these FS are used in a rule-based fuzzy logic system (FLS), the result is an interval T2 FLS (IT2 FLS). There are more sets of rule outputs executed in these FLSs than IT2 FS, and moving from these sets to numbers requires doing two successive actions: type reduction and deferentialization, as is usually required by most engineering applications in FLS. Type reduction maps the output T2 FS to type 1 (T1) FS, and defuzzification converts that T1 FS to a number. The type-reduction method was matured by Karnik and Mendel. If it is unromantic to an unqualified T2 FS, the calculation of an astronomical number is required. If you are unromantic to IT2 FS, you will need a very small number of calculations. This is one of the main reasons why IT2 FLS is sustentation, but unqualified T2 FLS is not.

In the case of the IT2 FLS, there may also be different kinds of type reductions. Everything that has been aged so far gives T2 FS a T1 center number, so when all the sources of uncertainty disappear, the output of IT2 FLS is reduced to that of T1 FLS. So calculating the center of IT2 FS plays an internal role in IT2 FLS. In addition, the heart of IT2 FS provides a measure of the uncertainty of IT2 FS, and has recently become an infrastructure for moving from data serenity to the next in more and more subject groups (the meaning of words and the intervals they socialize).

6.5 Type Reduction: Unspecified Consequences

The deepening of a T2 FS is a two-step process consisting of a first type reduction followed by a resulting second set of type 1.

The final step in a fuzzy reasoning system (FIS) is deepening. In a type 2 fuzzy reasoning system, differing consists of two stages: type reduction, which is a procedure of converting a set of type 2 to a set of type 1, and a suitable diffusing, which rewards sharp numbers by diffusing this type 1 set.

This type-reduction algorithm was originally described by Mendel with respect to a generalized T2 FS:

1. All possible type 2 embedded sets are listed.
2. There is a minimum secondary membership level for each embedded set.
3. For each embedded set, the type 1 centric domain value of the type 2 embedded set is calculated.

4. For each built-in set, the quadratic rank is paired with the domain values to create a series of ordered pairs (x, z). For some values of x, it can increase above each z value.

5. For each domain value, the maximum secondary grade is selected. This produces a set of pairs $(x, zMax)$ that are ordered so that there is a one-to-one correspondence between x and $zMax$. This completes the type reduction from the type 2 set to the type 1 set.

This strategy involves processing all embedded sets within a type 2 set. There are a lot of embedded sets, often totaling millions. Although individually hand-treated, the set included in the overall reward simply leads to a processing stick-up thanks to the higher cardinality. As a result, thorough deepening should be considered a theoretical flow rather than a practical skill. In woody discretization, this strategy can be implemented, but it is very slow. Preferably in discretization, the problem of memory space and a very large number of representations makes implementation impossible. Despite its practical shortcomings, we consider a thorough deferring algorithm as the standard for evaluating the validation of other type 2 deferring algorithms.

In the interval sample, all secondary grades are 1, which is how the minimum secondary grade is 1. So steps 2 and 5 can be removed from the spacing algorithm. All sorted pairs are of the form $(x; 1)$. Graphically they lie on the horizon.

6.5.1 Center Type Reduction

The centroid defuzzifier combines a set of output types 1 using t-conorm and then finds the center of this set.

6.5.2 Height Type Reduction

The height defuzzifier replaces each rule output set with a singleton at the point with the maximum membership in that output set and then computes the centroid of the type 1 set consisting of these singletons.

6.5.3 Set Center Type Reduction

The set-centric type reducer replaces each result set with a centroid (if the result set is type 2, it is itself a type 1 set) and finds weight stereotypes around these centroids.

6.5.4 Computational Complexity of Type Reduction

Type reduction was suggested by Karnik and Mendel and others. This is the "extended version" ("extended version" of the type 1 defuzzification method), and it is a tapped type reduction given that this moves from the FLS's type 2 output set to a type 1 set with tabs. "Then we can defer this set to get a single, well-finished number." The type-reduction set can be increasingly important than a single well-performed number because it conveys a measure of the uncertainty blown through the type 2 FLS. There are many kinds of type reductions such as set center, height, and modified height.

We have introduced several types of reduction methods. Unfortunately, they have high computational complexity. Fortunately, however, type 2 FLS can be thought of as a confusion of a large number of type 1 FLS. The type-reduced set is the sum of the outputs of all these embedded type 1 FLSs. Operations for each type 1 FLS can be processed in parallel. Therefore, the computational complexity for each type 1 FLS is the same for each deferential operation. The number of parallel processors depends on the number of type 1 FLSs required for a particular type-reduction method, which depends on the underlying membership and the sampling rate of the output domain. Calculation shows the computational complexity in terms of multiplication, addition, and division. Also, the calculation compensates for the number of operations for t-norm (Kawaguchi et al., 1993).

6.5.5 Conclusion

Let's consider a well-known cart-centric problem for system evaluation. In this particular problem, the cart can only move along a line in a frictionless plane, and the goal is to propel the cart from the specified initial position of the line to the middle position of the line, which forms a general rental problem. The controller's inputs for this problem are the current position coordinates of cart x and the current speed of cart v. The output of this fuzzy model is the gravity F, which should be non-romantic on the cart. The area of cart position x was limited from −0.75 to 0.75 m. The range of the cart speed v was limited from −0.75 to 0.75 m/s. The output gravity F was specified between −0.18 and 0.18 N, and the sampling time used was t=0.1 s. This set of parameters and constraints was further utilized in this experiment reported here.

Based on the problem described above, a zero-order IT2 TSK fuzzy model was designed. Specifically, it mimics all domains of the input variables x and v using five linguistic values represented by the IT2 fuzzy set. The input variables x and v are negative large (NL), negative small (NS), zero (0), and positive small (PS). To evaluate the proposed IT2 TSK assistant operating on a sparse rule basis, we simulated a lack of information by manually removing two fuzzy sets from each input domain. Considering the initial state of cart x=0.5 m and v=0.5 m/s, the existing IT2 TSK assistant and the proposed IT2 TKS are rational in this experiment using the Dumbo and sparse rule base if applicable for system performance comparison. It's romantic.

This experiment shows that the proposed IT2 TSK assistant is worldly wise to use a Dumbo or sparse rule base to produce reasonable results. The IT2 TSK with the proposed Dumbo rule wiring took less time to convert the wagon from the initial position to the target position with relatively smooth control compared to the performance of the existing TSK tide channel based on the Dumbo rule base. This means that the proposed IT2 TSK system surpasses the traditional IT2 TSK approach when using Dumbo rule wiring. Also interestingly, the proposed tideway took longer to increase the direction of travel, which may be useful in real-world rentals for the greatest dynamic stability.

Moreover, the IT2 TSK successfully pushed the cart to the target position with a relatively smooth curve, but the convergence time using sparse rule wiring in the proposed IT2 TSK was longer than when using Dumbo rule wiring in both approaches. However, considering the size of the rule base utilized, the proposed tideway can solve the same tenancy problem with only nine rules, whereas the existing approach requires 25 Dumbo rules. This arguably demonstrates the power of the proposed system in reducing system complexity.

This experiment reveals that the proposed IT2 TSK tideway is worldly wise to generate reasonable results using either a Dumbo or sparse rule base. The proposed IT2 TSK with the Dumbo rule wiring took less time to transform the cart from the initial position to the goal position with relatively smooth control, compared to the performance from the conventional TSK tideway based on the undertow of the Dumbo rule base. This indicates that the proposed IT2 TSK system outperforms the conventional IT2 TSK method when the Dumbo rule wiring is used. Moreover interestingly, the proposed tideway took longer to transpire the moving direction, which might be useful in real-world tenancy for the largest dynamic stability.

With the help of IT2 TSK one can successfully move the cart to the goal position with a relatively smooth curve, although the convergence time taken by the proposed IT2 TSK with sparse rule wiring was longer than those with the Dumbo rule wiring by either approaches. However, if the size of utilized rule bases are taken into account, the proposed tideway can solve the same tenancy problem with only 9 rules, whereas the conventional approach requires a Dumbo rule with 25 rules. This unmistakably demonstrates the power of the proposed system in system complexity reduction.

6.6 Exercise

1. Distinguish between type reduction and defuzzification.
2. What is the rationalization of computational complexity in type reduction?
3. Provide an example to show how complexity in type reduction can be simplified.
4. What are the variegated methods of type reduction?

References

Greenfield, Sarah, and Francisco Chiclana. "Defuzzification of the discretised generalised type-2 fuzzy set: Experimental evaluation." *Information Sciences* 244 (2013): 1–25.

Kawaguchi, Mayuka F., and Tsutomu Da-Te. "A calculation method for solving fuzzy arithmetic equations with triangular norms." In *Proceedings 1993 Second IEEE International Conference on Fuzzy Systems*, IEEE, Tokyo (1993), pp. 470–476.

Liu, Feilong. "An efficient centroid type-reduction strategy for general type-2 fuzzy logic system." *Information Sciences* 178, no. 9 (2008): 2224–2236.

Liu, Xinwang. "Efficient centroid computation of general type-2 fuzzy sets with linear secondary membership function." In *2011 IEEE International Conference on Fuzzy Systems (FUZZ-IEEE 2011)*, IEEE, Taipei, Taiwan (2011), pp. 2163–2169.

Liu, Xinwang, and Jerry M. Mendel. "Connect Karnik-Mendel algorithms to root-finding for computing the centroid of an interval type-2 fuzzy set." *IEEE Transactions on Fuzzy Systems* 19, no. 4 (2011): 652–665.

Mendel, Jerry M., and Feilong Liu. "Super-exponential convergence of the Karnik–Mendel algorithms for computing the centroid of an interval type-2 fuzzy set." *IEEE Transactions on Fuzzy Systems* 15, no. 2 (2007): 309–320.

Torshizi, Abolfazl Doostparast, Mohammad Hossein Fazel Zarandi, and Hamzeh Zakeri. "On type-reduction of type-2 fuzzy sets: A review." *Applied Soft Computing* 27 (2015): 614–627.

7

Modeling of Sustainability

7.1 The Meaning of Sustainability

The biological and physical environment provides the economy with resources (such as water, air, fuel, food, metals, minerals, drugs, cycles of H_2O, C, CO_2, N, and O_2), photosynthesis, and soil formation.

According to the struggle to monetize all three global services, the total value in 1997 was $33 trillion per year. These numbers give you an idea of the importance of the environment.

Economic growth is based on these three services, and economic growth cannot be extended indefinitely because the global ecosystem does not grow. It is the personal job of many economists to solve the finiteness problem of ecosystems by replacing one form of want with a flip side. However, there are limitations to substitution, and often it cannot be said due to the scale of the global ecosystem or the laws of nature. There is no substitute for a stable climate, and no substitute for extinct species and services to nature.

In the past decades, we have witnessed dramatic environmental changes as a result of human economic activities. Global warming, stratospheric ozone depletion, species extinction, and a vacuum of natural resources provide a strong evidence that the current trivial phenomena cannot last longer. As a result, there is a growing interest in the concept of sustainability and sustainable development among policy makers and scientists.

The sustainable detail is not necessarily an intermediate growth, but it is the revival of various social sectors such as health, education, or environmental conditions. So what is sustainable development?

According to Brundtland, in increasingly undefined terms, the sustainable detail is

> Development that meets current needs without compromising values for future generations to meet their needs.

According to International Union for the Conservation of Nature (IUCN), it is

> Development to improve the quality of human life while conveying the theme of ecosystem support.

In general, sustainability incorporates many aspects of the environment, economy, and society. It also covers a variety of geographic sizes such as ecosystems, regions, countries, and the world. Moreover, it has a limited time range. Sustainability includes long-term goals but cannot be planned for thousands of years in the future. Always we make adaptive decisions that bring the tropics to our goals as much as possible.

7.2 Introduction to Sustainability by Fuzzy Assessment (SAFE) Model

Sustainability by Fuzzy Assessment (SAFE) is a hierarchical fuzzy inference system. Using "IF–THEN" rules and knowledge encoded with fuzzy logic, we combine 75 input, tab-delimited critical indicators into increasingly complex variables that describe different environmental and social aspects, and finally [0, 1].

This model was introduced by Phillis and Andriantiatsaholiniaina (2001), Andriantiatsaholiniaina et al. (2004), and Kouloumpis et al. (2008).

7.3 SAFE Model Overview

Overall sustainability (OSUS) is a combination of two main components: ecological sustainability (ECOS), and social or human sustainability (HUMS).

Ecological inputs consist of four secondary components:

1. Land integrity (LAND),
2. Water quality (WATER),
3. Air quality (AIR)
4. Biodiversity (BIOD)

The human-level components of sustainability are as follows:

5. Political aspect (POLICY)
6. Economic welfare (WEALTH)
7. Health (HEALTH)
8. Education (KNOW)

Each secondary component is assessed using the pressure-state-response flow of the Organization for Economic Cooperation, which assumes that humans put pressure on the environment (state) and are undeniable for irrefutable responses.

It can be obtained by combining essential indicators that cannot be removed. For example, the Indicator PR (BIOD) uses six essential indicators to reward the proportion of all endangered (or vulnerable to endangered) species (mammals, birds, plants, fish, reptiles, and amphibians). It measures the pressure on diversity.

The data processing sequence is as follows:

- Collection and normalization of misogynism data in [0, 1]
- Exponential smoothing
- Enter missing data
- Fuzzy traction of sustainability: fuzzy traction of essential indicators, traction of tertiary indicators (PR (LAND), ST (LAND), RE (LAND),...), traction of secondary indicators (LAND,..., POLICY,...), ecological and human components (HUMS, ECOS), and the traction of overall sustainability (OSUS)

7.4 Key Indicators of Sustainable Development

A total of 75 mandatory indicators are used in 128 countries. The definitions of these indicators are provided, Organization for Supply and Agriculture (FAO), United Nations for Conservation of Nature (IUCN), Organization for Economic Cooperation and Details (OECD, website), United Nations Environment Program (UNEP), United Nations Department of Statistics (2006), United Nations Detailed Program (UNDP), United Nations Organization for Education, Science and Culture (UNESCO), United Nations Framework Convention on Climate Increase (UNFCCC), and World Health Organization (WHO). In addition, many of these references provide annual data close to the indicators that are of little importance for most countries around the world.

7.5 Measuring Sustainability

There is no universal definition of sustainability. Values and political and economic interests play an internal role in the sustainability debate. However, from a scientific point of view, an irrefutable approach provides a tool for comparison between countries that have documented their path towards sustainable development.

An overview of this approach is provided in the following.

Pressure state response (PSR)

This model was completed by the Organization for Economic Cooperation and is based on the fact that humans put pressure on the ecosystems and societies in which they maintain their status and are undeniable for their undeniable responses. Socio-economic indicators are attracting more attention, but the main focus is on the ecological side.

Ecological footprint

Introduced by Rees, it calculates the equivalent land required to produce essential non-removable resources and churn the non-removable waste associated with a given population. Simply put, an ecological footprint is a productive land used by the population. It is unreasonable in terms of ecology, and it calculates land area, not sustainability score.

Indicator of sustainability

This model was introduced by IUCN and is a visual tool for sustainability assessment. There are two basic components of a country's sustainability: ecosystem well-being and human well-being. All metrics are scaled by [0, 100], where 0 is the worst performance and 100 is the more weighted performance of the metric. Then, the score is calculated as a simple count.

Environmental sustainability tabulate (ESI)

ESI calculates a country's environmental sustainability table based on 21 indicators, which in turn are evaluated on 76 datasets. The ESI table is calculated as a weighted stereotype of metrics with the same weight. The country ranking is determined accordingly.

Driving sustainability by fuzzy assessment (SAFE)

This model was introduced by Phillis and Andriantiatsaholiniaina (2001), Andriantiatsaholiniaina et al. (2004), Kouloumpis et al. (2008), and Phillis and Kouikoglou (2009). SAFE is a hierarchical fuzzy inference system. Using "IF–THEN" rules and knowledge encoded with fuzzy logic, we combine 75 input, tab-delimited critical indicators into increasingly complex variables that describe different environmental and social aspects, and finally [0, 1].

Multiple criteria and fuzzy logic

An ESI-like model using 74 indicators and multi-criteria decision-making (MCDM), along with a fuzzy inference scheme similar to SAFE, was introduced by Liu. While MCDM has three stages: decomposition, weighting, and synthesis, it computes a volume sustainability table through sequential fuzzy inference.

Sustainable society table (SSI)

SSI is based on 22 environmental and social indicators aggregated into five main categories using equal weights. Then, the five categories are aggregated into SSI using unequal weights. It is ranked accordingly in all 150 countries.

7.6 Fuzzy Assessment

Sustainable decision-making involves complex and often poorly defined parameters with an upper layer of uncertainty due to an incomplete understanding of the underlying problem. The dynamics of social environmental systems cannot be explained with traditional mathematics, taking into account the inherent complexity and ambiguity. Also, the concept of sustainability is polymorphic and full of subjectivity. Therefore, it is getting closer and closer to using fuzzy logic for evaluation.

Fuzzy logic is a scientific tool that allows you to model systems without detailed mathematical explanations using qualitative and quantitative data. The SAFE model uses fuzzy logic to calculate complex indicators (outputs) from critical items (inputs). Calculations are washed into words using knowledge expressed in formal language rules.

```
if
        ( inputs )
then
        ( outputs )
```

Two examples of such "IF–THEN" rules, which are used in the first and last stages of the SAFE inference process, are given in the following.

Assessing a tertiary variable from vital indicators:

```
if
        'Threatened Mammals' is Medium
        'Threatened Birds' is Strong
        and 'Threatened Plants' is Medium
        and 'Threatened Fishes' is Weak
        and 'Threatened Reptiles' is Strong
        and 'Threatened Amphibians' is Strong
then
        PR(BIOD) is Bad.
```

Assessing OSUS from its primary components:

```
if
        ECOS is Bad
        and HUMS is Good
```

```
then
        OSUS is Intermediate.
```

The terms Medium, Bad, Intermediate, etc. in the rules given whilom represent fuzzy sets. Each rule has a given stratum of truth or firing strength, which is a volume measure of the stratum to which its inputs vest to the respective fuzzy sets. The method we describe next, tabbed fuzzification, is used to compute the stratum to which a vital indicator belongs to a specific fuzzy set.

The normalized vital indicators are fuzzified using three fuzzy sets with linguistic values:

- Weak (W)
- Medium (M)
- Strong (S)

For composite indicators (primary, secondary, and tertiary components), five linguistic values are used:

- Very Bad (VB)
- Bad (B)
- Average (A)
- Good (G)
- Very Good (VG)

The overall sustainability is measured using nine fuzzy sets:

- Extremely Low (EL)
- Very Low (VL)
- Low (L)
- Fairly Low (FL)
- Intermediate (I)
- Fairly Upper (FH)
- High (H)
- Very Upper (VH)
- Extremely Upper (EH)

Each indicator value x belongs to one or increasingly fuzzy sets with unrepealable membership grades. For simplicity, triangular membership functions $\mu(x)$ are used. For example, in 2002, 13.7% of the mammal species in Greece were endangered. The target value for this indicator is $T=\tau=0\%$, and the upper threshold of unsustainable values is $U=35.5\%$. The normalized value for this indicator is $x=(13.7-35.5)/(0-35.5)=0.614$. This value belongs to the fuzzy set Medium with membership grade $\mu M(0.614)=0.965$ and to the fuzzy set Strong with grade $\mu S(0.614)=0.035$.

7.7 Sensitivity Analysis

Sensitivity wringer plays a fundamental role in the creation of visualizations given that the visualization parameters determine the effect of transpiration on system performance. In this section, we strive to provide a pun to the question of how to diamond a policy for sustainable development. SAFE models can help visualization creators formulate sustainable policies by evaluating sustainability for a variety of development scenarios. Scenarios are specified by indicators of misogynistic sustainability, which mainly reflect the consequences of policies and actions taken in a specific period. Reverting these values and observing changes in sustainability as a result can identify the most important indicators contributing to sustainable development. This procedure is called sensitivity analysis.

Sensitivity wringer entails calculating slopes of ECOS, HUMS, and OSUS with respect to each required indicator. The slope provides an increase in sustainability per unit increase in some important indicators. A sensitivity test can be performed using the following steps:

1. Calculation of OSUS:

    a. For a given country, normalize and smooth all indicator data using the methods described in previous sections.

    b. Fuzzify the vital inputs.

    c. Compute the membership grades of composite indicators to the fuzzy sets VB, B, A, G, and VG. Start from the inference engines that use only vital indicators as inputs and proceed successively to the ones that use composite indicators. Finally, compute the membership grades of OSUS to the nine fuzzy sets EL, VL, ..., EH and compute a well-done value for OSUS by height defuzzification.

2. Introduction of perturbation:
    For some vital indicators, say, c increases its normalized value $xc \in [0, 1]$ by some stock-still value δ, for example, 0.1 or 10%.
    If the result is greater than one, then truncate it to one to stave overshooting permissible regions of indicators.

3. Sensitivity analysis:
    Assess the overall sustainability using the same set of data as in step 1 except for indicator c whose value is now $xc\ \delta$.
    Denote the new towage by OSUS($xc\ \delta$). The gradient of OSUS with respect to xc is specified by the forward difference:

$$\Delta c = OSUS(xc\ \delta)OSUS$$

Reset the vital indicator c to its original value xc.

4. Loop:
   Repeat steps 2 and 3 for all vital indicators.
5. Ranking:
   Identify the gradients with the largest values, which correspond to the vital indicators that stupefy OSUS the most.

An important full-length of the SAFE model is monotonicity. Whenever a vital indicator of sustainability is improved, the components of sustainability that depend on this indicator as well as OSUS increase or at least do not decrease, that is, if $\delta \geq 0$, then $\Delta c \geq 0$. The use of product-sum algebra in all inference engines ensures that the hierarchical fuzzy system is monotonic.

By waffly several indicators simultaneously in step 3, we can compute gradients of higher orders and formulate increasingly comprehensive environmental policies. For example, the second-order gradient of OSUS with respect to indicators c and c' is

$$\Delta c,c' = OSUS(xc\ \delta, xc'\ \delta)OSUS$$

Sensitivity wringer is unjust towards indicators which vest to small groups. For example, RE(AIR) depends only on production of renewable resources. Therefore, an increase in the latter directly affects the former. ST(AIR), on the other hand, depends on four vital indicators. An resurgence of one of these indicators will result in a small resurgence of ST(AIR). To stave this bias, a vital indicator c is ranked equal to the product:

$$Dc = (1 - xc)\Delta c$$

where $1 - xc$ is the loftiness of indicator c from the sustainable value, and Δc is the gradient of OSUS with respect to xc. Thus, those indicators that stupefy OSUS the most and are farther in the unsustainable region are pinpointed and ranked accordingly.

7.8 Advantages and Disadvantages of the SAFE Model

SAFE uses fuzzy logic that does not require an explicit mathematical model to sum up volume indicators and can process both quantitative and qualitative information. Fuzzy logic often prevents the use of weights that are ill-headed or cannot be extracted by the creator of the visualization. Moreover, SAFE is a rather simplistic model that respects the non-rewarding attribute and is the only flow that evaluates sustainability taking into account the value in the time dimension.

SAFE, on the other hand, has obvious drawbacks that can also be found in other models.

F: It is somewhat subjective and does not have a mechanism to limit the number of inputs to what is actually needed. There is an irreversible overlap in the display period. For example, the number of hospital beds overlaps with public health expenditures or city gross particulates, and urban NO_2 concentrations overlap with mortality from respiratory diseases. However, finding a causal model linking these indicators is indescribable.

Ratio: The rule base and membership features reflect the values, knowledge, and prejudices of those who invented them. SAFE's rule base attaches the same importance to input variables as they are organized in a different team way without consulting an expert. The subjectivity of modeling is not surprising given that sustainability is not a concept responsible for a strict definition. This is a sample with all other sustainability models.

Seed: More work remains to improve the weighting and membership functions of non-delete indicators such as CO_2 emissions, nuclear and hazardous waste, biodiversity loss, government debt internals, etc. to capture new sustainability issues as reality changes.

7.9 Sample Study for the SAFE Model

7.9.1 SAFE for Energy Sustainability

It cannot be supported that a country's overall energy sustainability (OSUS) is a combination of two major components: ecological sustainability (ECOS) and human sustainability (HUMS). Ecological inputs consist of two auxiliary elements: air quality (AIR) and soil quality (LAND). The human component of energy sustainability is social (accessibility) and economic (production, consumption, security). Each secondary component is evaluated using a mandatory, non-delectable indicator that is an input to the system.

The order of data processing is similar to the original SAFE model.

- Misogynist data collection
- Normalization of [0, 1]
- Exponential smoothing
- Data replacement
- Fuzzy
- Fuzzy drive of sustainability
- Create sensitivity tweaks and visualizations

7.10 Conclusion

Sustainability is the process of maintaining transpiration in a well-transformed environment in which the exploitation of resources, investment direction, direction of technical details, and institutional escalation are all harmonized and can improve present and future potential to meet human needs. For many on the site, sustainability is specified through interconnected domains or pillars. Sub-areas of the sustainable sub-area were also considered: culture, technology, and politics. Sustainable detail may be an organizational principle for sustainability for some, but for others, the two terms are paradoxical (i.e., detail is inherently unsustainable). Sustainable detail is the detail that meets the needs of today without compromising the value that future generations will meet their needs. The Brundtland report for the World Environment and Details Committee introduced the term "sustainable development."

Moreover, sustainability can be defined as a socio-ecological process characterized by the pursuit of a global ideal (Azar et al., 1996). However, by constantly and dynamically becoming irreversible, the process leads to a sustainable system.

Healthy ecosystems and environments are necessary for the survival of humans and other organisms. The ways to reduce the negative human impacts are environmentally friendly chemical engineering, environmental resource management, and environmental protection. Information comes from unattended computing, unattended chemistry, earth science, environmental science, and conservation biology.

Moreover, the move towards sustainability is a social move that entails international and national laws, urban planning and transport, integrated supply management, regional and individual lifestyles, and virtuous consumerism. Increasingly sustainable lifestyles are reorganizing living conditions (e.g. eco villages, eco locals, and sustainable cities) and economic sectors (permanent cultivation, untried buildings, sustainable agriculture) or working practices (sustainable architecture). It can take many forms, including re-evaluation. There is an urgency for science of developing new technologies (green technology, renewable energy, and sustainable fission and fusion power) or designing systems in a flexible and reversible manner and coordinating individual lifestyles conserving natural resources (Gilg et al., 2005).

"The term 'sustainability' should be viewed as humanity's target goal of human ecosystem equilibrium (homeostasis), and 'sustainable development' refers to the whole flow and temporal process leading us to the end of sustainability." Despite the increasing use of the term "sustainability," the possibility of human society decompressing environmental sustainability continues to be questioned in terms of environmental destruction, climate change, overconsumption, population growth, and social pursuit.

7.11 Exercise

1. What is sustainability?
2. Provide a short description of SAFE model.
3. What is the primary reason that SAFE model is quite successful in measurement of sustainability?
4. Describe hierarchical structure of the SAFE model with the help of a labeled diagram.
5. How can energy sustainability be measured by SAFE model?

References

Andriantiatsaholiniaina, Luc A., Vassilis S. Kouikoglou, and Yannis A. Phillis. "Evaluating strategies for sustainable development: Fuzzy logic reasoning and sensitivity analysis." *Ecological Economics* 48, no. 2 (2004): 149–172.

Azar, Christian, John Holmberg, and Kristian Lindgren. "Socio-ecological indicators for sustainability." *Ecological Economics* 18, no. 2 (1996): 89–112.

Gilg, Andrew, Stewart Barr, and Nicholas Ford. "Green consumption or sustainable lifestyles? Identifying the sustainable consumer." *Futures* 37, no. 6 (2005): 481–504.

Kouloumpis, Viktor D., Vassilis S. Kouikoglou, and Yannis A. Phillis. Sustainability assessment of nations and related decision making using fuzzy logic. *IEEE Systems Journal* 2, no. 2 (2008): 224–236.

Phillis, Yannis A., and Luc A. Andriantiatsaholiniaina. "Sustainability: An ill-defined concept and its assessment using fuzzy logic." *Ecological Economics* 37, no. 3 (2001): 435–456.

8

Epilogue

8.1 Introduction

Fuzzy set theory was first introduced by Zadeh in 1965. Fuzzy sets and systems have been evolving for over 50 years and have been chosen as the methodology of tower systems that can release satisfactory performance in situations of scattered uncertainty and inaccuracies. Therefore, fuzzy logic system (FLS) has been successfully implemented in many practical applications including modeling and control, time series prediction (TSP), and data mining.

The type 1 fuzzy logic system (T1 FLS) is the most well-known and widely used type of FLS. Nevertheless, in recent years, there is a significant increase in research into more and more forms of fuzzy logic, such as the interval type 2 fuzzy logic system (IT2 FLS) and more recently undefined type 2 FLS (T2 FLS). This transition motivated me to realize that type 1 fuzzy sets (T1 FS) can only handle a limited level of uncertainty, while real-world applications often face multiple sources and higher levels of uncertainty. In 1975, Zadeh recognized these potential limitations and introduced the concept of a (generic) type 2 fuzzy set (T2 FS), an extension of the T1 FS. As an increasingly pervasive model, the T2 FS is considered to be potentially best suited for modeling uncertainty. The extra complexity arises including footprint of uncertainty (FOU) and 3D, and provides a level of self-domination for T2 FS compared to T1 FS. However, due to the same complexity, FLS with T2 FS is computationally very burdensome compared to FLS with T1 FS.

The computational complexity of using T2 FLS has led to the introduction of the simplified IT2 FLS, the most surprisingly used type of T2 FLS today. The IT2 FLS uses IT2 FS, a special sample of unqualified T2 FS with all secondary membership levels of 1 (Nie et al., 2008). Many researchers prefer IT2 FLS over T1 FLS due to its potential to model and mitigate the effects of uncertainty. From this, it is important to note that many of the research works considering T2 FS in the study do not quantify the "high" method, but they expect diamonds to perform well in the "high level of uncertainty" squatter.

So, the main problem with the use of FLS is the type of fuzzy set and its parameter and interpretation of parameters such as the number of rules, but an increasingly fundamental problem is whether you should use T1 or T2 FS. Although there are experimental traces of improvement in terms of handling uncertainty in IT2 compared to its T1 counterpart, no systematic method has been developed to determine the potential soundness of using T2 FS over T1 FS.

TSP can be used as a well-defined sample application (Aladi et al., 2014) to obtain a usage-led survey of the relationship between the FOU size of FS and the level of uncertainty in use. This can be done by diamond marking the T1 FLS for the TSP and creating a variety of FLSs that continuously increase the FOU size over several steps. At the same time, we inject increasing levels of noise to provide a priori known and well-defined/understood source of uncertainty. Hence, using time series prediction provides a platform to explore the behavior of FLS with different FOU sizes for different levels of noise/uncertainty. The main goal is not to decompress the optimal performance in the predictions, but to illuminate the approaching size of the FOU for a given level of uncertainty/noise. The results are obvious to use, but they provide important insights into the relationship between FOU size and level/uncertainty in a given setup.

8.2 Type 2 vs. Type 1 FLS

Type 2 fuzzy sets may represent uncertainty in the membership capabilities of the T1 FS. So, in the case of a type 1 fuzzy system, the input data contains noise, so this uncertainty can be transferred to the membership function Uncertainty and the type 2 fuzzy system can be rebuilt. The type 2 fuzzy system can handle the uncertainty of rules, system parameters, and system inputs. Thus, it creates a powerful extension of an important type 1 fuzzy system.

8.3 Application for Type 2 FLS

There is an intrinsic and unique relationship between fuzzy sets, probabilities, and statistics that have been recognized since fuzzy sets were proposed. Statistical theory is based on probability theory. Thus, fuzzy statistics can take many forms depending on which probability (uncertainty, spacing, fuzzy) theory is used.

The details of fuzzy probability theory began with the introduction of fuzzy random variables in 1978 by H. Kwakernaak. Fuzzy probabilities represent a special sample of inaccurate probabilities associated with the concept of a random set.

Since fuzzy events were proposed by Zadeh, using higher-order fuzzy events to present real-value problems is actually an inhibitor of fuzzy probabilities. Traditionally, the α-level set representation should be a convex and normalized fuzzy set of fuzzy random variables for fuzzy events. However, real fuzzy events cannot invariably meet these requirements, which can suppress the details of real fuzzy probabilities. "Type 2 fuzzy event J-plane" is specified to represent concave and denormalized type 2 fuzzy events, as well as convex and normalized events, and a new concept of membership probability density function (Mpdf) is presented. Following the "Type 2 fuzzy probabilistic system," we acquired the type 2 fuzzy SARIMA system for seasonal time series prediction in which real numerical uncertainty and abnormal data are intensive to realize a type 2 FLS that evolves from a rule-based fuzzy system in a system based on a type 2 fuzzy probability model.

8.4 Rule-Based Nomenclature for Video Traffic

Video-based surveillance systems have a wide range of applications for traffic monitoring because they provide more and more information compared to other sensors. A rule-based framework for policy and concern detection in traffic video can be obtained from stationary video cameras. Moving targets are segmented from the image and tracked in real time. They are classified into various categories using a new Bayesian network approach that uses image features and image sequence-based tracking results for robust classification. Trace and nomenclature results are used in a programmed context to understand the behavior. Two main types of interactions were considered for policy awareness. One is the interaction between two or increasingly mobile targets in the camera's field of view (FoV). The other is the interaction between an object and a stationary object in the environment. The framework is based on two types of prior information. It is a set of predefined policy scenarios that need to be analyzed in 1) camera FoV in relation to the various fixed objects in the scene and 2) in various situations. The system can recognize the policy in the video and reward the vocabulary output of the detected action. It can also handle the uncertainty caused by visual signal processing errors. Successful policy recognition results for pedestrian–vehicle interaction and vehicle–check post interaction can be demonstrated.

8.4.1 Selected Function

It can solve a lot of ambiguities, so working in world coordinates is bigger than image coordinates (Shotton et al., 2013). For example, using perspective shortening can erroneously perceive target motion in the image plane. If the ground speed is similar, it is flat. Camera parameters are required to convert a measure of image coordinates to a measure of world coordinates. This only needs to be cleaned once for camera setup and image coordinates. The world coordinate axis is selected so that the XY plane is aligned with the ground plane of the scene and the Z rotation is perpendicular to the ground plane. The perspective transformation equation for a pinhole camera model is calculated using manually selected points in image space and coordinates in world coordinate space.

At least six point matches are required to calculate the perspective projection matrix. Select each point more than this minimum number, use least squares to solve overly constrained linear equations, and filter out noise due to measurement errors. I used the standard dimensions of the road markings to get the world coordinate values. So, there is no need to make ground measurements on the road.

Using a Bayesian network, naming an object as one of its tracking classes (pedestrian, motorcycle, car, truck, heavy truck, and noise) can indicate the height of the object. This value was obtained by measuring the typical height of objects in various classes.

8.4.2 FOU on Function

In word counting, words can be modeled as spacing values (IV) fuzzy sets (FS). It was a very important matter to organize the FOU for FS in almost one word. Although many methods have been developed, it is still difficult for each person to provide a FOU close to their own words. The main reason is that the FOU is inherent in representing the likelihood of a variable value. To overcome this problem, we can aim to interpret the FOU of the FS by revealing the possibility of the variable value.

It turns out that the focal point of FS is a measure of the uncertainty inherent in the FOU. The normalized weights of the variable values used to calculate the centroid can be considered a type 1 (T1) membership function (MF) and are simple to show the likelihood of these values. Research can be conducted by uncovering the relationship between FOU and this T1 MF. This T1 FS is tabbed with T1 (ET1) FS equivalent to IV FS. You can set up an equation that relates the lower MF (LMF) and upper MF (UMF) of IV FS to the MF of ET1 FS. Using the established equations, we can present properties close to the MF of ET1 FS for IV FS, indicating the relationship between FOU and MF of ET1 FS. These properties help people relate the FOU of IV FS to the likelihood of a variable value that occurs.

8.4.3 Rules

Calculating the center of the IV FS was a very important matter, and many methods have been matured. The Karnik–Mendel (KM) method is not the best judged defuzzy method. However, this represents the endpoint of the central gap, which is important in this context. The KM method has been matured based on the wavy fragment representation theorem, which can be considered a combination of all T1 FS with IV FS (Mendel et al., 2007). You can start by calculating the centroid of the embedded T1 FS involved. The center of all embedded T1 FS is locally spaced by the left and right endpoints.

The FOU for IV FS is implied in showing the likelihood of a variable value, preventing people from providing FOUs close to their own words. The MF of ET1 FS for IV FS easily shows the possibility of variable values. Equations related to UMF and LMF of IV FS and MF of ET1 FS have been established. Properties representing the relationship between the FOU of IV FS and the MF of ET1 FS were presented using the established equation. These properties help people relate the FOU of IV FS to the likelihood of a variable value.

8.4.4 FOU for Measurement

As we know, "the word is the middle point of variegation for variegated people, so it's unclear. So you need an FS model for words that are likely to capture uncertainty, and you need to use Interval Type 2 (IT2) FS as your word model." Four approaches have been reported to build IV FS: a gap approach, an improved gap approach, a gap endpoint assistant, and an MF approach.

The Interval Tide Way allows each person to provide an endpoint of close spacing to his or her horse. Data intervals for all subjects are mapped to T1 MF, and then these type 1 (T1) MF (membership functions) are aggregated to form an FOU. The enhanced spacing assistant is similar to the spacing approach, except that the previous method preprocesses calm data and provides an increasingly rational procedure to obtain a low MF. Also, the gap endpoint assistant allows each person to provide an endpoint of the gap close to the person's speech, but it gets FOU in a variety of ways. In a calm data interval, the endpoint statistic is set and converted into an uncertainty measure of IV FS. Finally, you can set up a parametric FOU so that the measure of uncertainty matches the extracted data. It has been proven that the center of IV FS can be used as a measure of the uncertainty inherent in the FOU. As a measure of uncertainty, you can implement gap endpoint tides that use the center of the IV FS.

In the personal MF approach, each person cannot provide an FOU close to his or her words, and the final FOU can be performed by aggregating the FOUs of various people. However, it is still a difficult task for each person to provide an FOU. The main reason is that the FOU is inherent in representing the likelihood of a variable value, and people are inflexible to relate to each other. The centerpiece of IV FS has been proven as a measure of the uncertainty inherent in the FOU. The normalized weights of the values of the primary variable used to calculate the centroid are simple to show the likelihood of the variable values. These normalized weights can be considered T1 MF. Research can be conducted by uncovering the relationship between FOU and this T1 MF. For convenience of explanation, this T1 FS is tabbed as T1 (ET1) FS equivalent to IV FS. Of the conventional methods matured to calculate the center of the IV FS, the KM method has been widely used. You can set up an equation that relates the lower MF (LMF) and upper MF (UMF) of IV FS to the MF of ET1 FS. Using the established equations, we can present properties close to the MF of ET1 FS to represent the relationship between the FOU of IV FS and the MF of ET1 FS.

8.4.5 FL RBC's Parameters

FLS is one of the common approaches to soft computing methods. Since its first visit in the 1970s, it has been unromantic in many areas. As a formal description, a fuzzy logic system (or fuzzy system) is a computing system in which a fuzzy set is used for the source and/or the result of a fuzzy IF–THEN rule. Likewise, the attractiveness and success of FLS comes from the core features it pursues:

- Ability to flexibly represent objects with fuzzy sets
- Interpretability of results by providing insight into the visualization production process.

J. Mendel proposed a new direction for fuzzy system research that considers type 2 fuzzy sets (T2 FS) in rule wiring instead of using the usual fuzzy set (type 1 fuzzy set – T1 FS). By definition, T2FS is a fuzzy set, and the membership level of each element is T1FS of [0,1]. Therefore, the membership function of T2 FS is three-dimensional. The benefit of this flow is theoretically reducing the impact of uncertainty on fuzzy logic systems. In particular, J. Mendel emphasized the health of T2 FS in that it is worth dealing with potential noise in the training data. There has been a lot of works in the past decade on T2 FS and T2 FLS both in theoretical and practical terms. Some of the main approaches are as follows:

- First, reinforce the basic theory towards a well-formed T2 FLS with computational skills for T2 FS such as FOU concept, expression theorem, set theory operation, centroid, inference, and type reduction.
- Second, apply T2 FS and T2 FLS to perceptual modeling, perceptual computing, and word computing.
- Finally, in areas such as control and medicine, training data demonstrates the effectiveness of T2 FLS in problems such as noisy or uncertain predictions, and classifications.

Recently, a new entrance ceremony of type 2 fuzzy set-hedging trigonometry type 2 fuzzy set (HaT2FS) has been proposed. Apparently, HaT2FS is a fuzzy set with language membership levels. However, the flow of HaT2FS varies fundamentally from study to study. Instead of traditional processing, HaT2FS's numbering and reasoning works based on the nature of the linguistic truth values of hedging algebra.

To prove the applicability of the hedge trigonometric type 2 fuzzy logic system (HaT2-FLS), and to take enough of the HaT2FS sound battleship, you can apply the experimental method after applying the method of constructing HaT2-FLS from input and output data. Among knowledge-based systems research, the pattern naming problem is one of the most important things. The key to this problem is that there is a possibility that there will be uncertainty in the data (i.e. the data contains noise). There are many different approaches to deal with this problem, from statistical-based techniques such as the subconscious Markov model to soft computing methods such as tense neural networks and fuzzy rules-based naming systems. Among them, the contribution of fuzzy logic systems to the problem is highly appreciated.

8.4.6 Calculation Formula for Type 1 FL RBC

Knowledge-based systems can capture and store expertise and drive reasoning knowledge through reasoning techniques. Expert knowledge refers to what an expert has experienced or is known to be true. Reasoning knowledge refers to new knowledge derived through the process of reasoning. You can capture and express your expertise in a variety of ways to express your knowledge. One of the most surprisingly used methods is to use the fuzzy production rules proposed by Negoita and Chen. The fuzzy production rule is a rule that describes the fuzzy causal relationship between the source part and the result part of the rule, and can induce reasoning knowledge through various knowledge reasoning methods.

Knowledge reasoning determines whether there is a pre-consequence relationship P Q. That is, if the knowledge of P is true, then the knowledge of Q is true. In spontaneous nomenclature issues, alternatives are judged in terms of irreversible criteria, such as determining a wall consumer's credit tier based on some (ambiguously explained) characteristics, such as "financial basis," "Personality," and "Security."

Knowledge of the real environment is not well expressed, but incomplete and inaccurate, or simply ambiguous. Many fuzzy production rules have been proposed for fuzzy knowledge representations to capture incomplete and inaccurate expertise. Fuzzy production rules derive from traditional production rules except that the proposition of the rule includes some linguistic terms instead of a word-by-word number measure. Fuzzy sets can be used to represent various linguistic terms (e.g. very well, tall, very heavy) of

a language variable (e.g. health, height, weight). Various interpretations or
shadings of linguistic terms can be expressed in the degree of truth or the
level of membership of the fuzzy set. Memberships range from full to non-
members, from 1 to 0.

8.4.7 Calculation Formula for Type 2 FL RBC

In recent decades, many approaches have been proposed for generating
fuzzy IF–THEN rules from numerical and nominal data. It's difficult to
reward well-formed surveys for creating/extracting fuzzy rules. The pursuit
is a non-persistent and incomplete summary of this subject.

- Fuzzy rules for nomenclature can be extracted from the data through
  a simple heuristic procedure. Here, heuristic information plays a key
  role in rule extraction.
- The derivation flow of a fuzzy visualization tree, first constructing
  a visualization tree and then transforming the tree into a fuzzy IF–
  THEN rule set, is important for fuzzy rule extraction.
- Fuzzy rules for nomenclature can be created using neurofuzzy
  techniques.

Let's consider an inference process that involves several parameters that
need to be determined. Regarding the facts given, the inference conclusion
will be a parametric fuzzy set, which means that the inference conclusion
will be confusing with various parameters. I prefer parametric fuzzy sets
with maximum fuzzy entropy (rather than other fuzzy sets) as inference
conclusions according to given constraints. Fuzzy entropy maximization
can be realized by parameter adjustment. Why can fuzzy entropy maximization modernize naming accuracy? The pursuit is an intuitive explanation. I have a naming problem for admission and assume A is the object to
be classified. In the absence of extra information about the nomenclature
available, the most reasonable nomenclature result for A should have the
same likelihood that A belongs to each of the n classes (i.e. 1/n), which
would be achieved by maximizing the entropy of A. If some preliminary
information about nomenclature is misogynistic (i.e., there is a training set
for which admission of each example is known), then you need to maximize the entropy of A to get a reasonable and pearly nomenclature for
A. Constraints indicate that training examples can be classified correctly.
These constraints are an intermediate point where misogynistic information about nomenclature has been utilized and the remaining uncertain
information about nomenclature is handled by maximum entropy. A reasonable and pearly nomenclature for A is expected to lead to improved precision. Since A is the leftover object to classify, you should well construct
A's entropy maximization by maximizing the entropy of the training set.

Unfortunately, so far we are unable to reciprocate the official mathematical formula for the previous explanation. It will be considered an important issue for remote research.

8.4.8 Optimization of Rule Design Parameters

You can use the Takagi–Sugeno fuzzy nomenclature system (T-S FCS) using particle cluster optimization (PSO) and the support vector machine (SVM) for parameter optimization. T-S FCS is synthesized by a fuzzy IF–THEN rule whose result is a linear equation of state. The guardian of T-S FCS is not shaken by the fuzzy membership of the input full-length vector. During the source configuration process, predefined values are optimized remotely using PSO. The resulting parameters of T-S FCS are learned through SVM. The proposed T-S FCS is worldly in minimizing the impact of uncertainty, reducing the influence of transforming factors, and repaying the system maximum generalization performance inheriting the advantages of T-S fuzzy systems, PSO, and SVM.

It is very important to set the initial values of the parameters, which directly reduces the quality of the algorithm. So, during the source configuration process, the values pre-specified are optimized remotely using PSO. PSO is an evolutionary computational technique and is similar to genetic algorithms (GA) in that the system is initialized with a set of random solutions. However, it differs from GA in that each potential solution is defined at a random rate. Thus, PSO outperforms GA in terms of search processing time and convergence speed.

8.4.9 FL RBC Test

In general, fuzzy tenancy systems are still the most important applications of fuzzy theory. This is a generalized form of expert tenancy that uses fuzzy sets with fuzzy rules in system modeling. An important idea in Zadeh's and Mamdani's classic fuzzy approaches is to summarize the conclusions by evaluating the concordance layer from observations that triggered one or several rules in the model. In most fuzzy modeling or fuzzy tenancy systems, experiments and simulations are set up to produce datasets that more weightily describe all possible outcomes. Without this, human experts create a set of more weighted fuzzy rules that do tenancy or modeling. In general, fuzzy rules created in this way weaken the entire universe, taking into account all possibilities. However, serious problems can arise due to the high computational time and spatial complexity of the rule base used to describe models with multiple input variables with adequate accuracy. "Exponential explosion" seldom allows unqualified systems that use classic fuzzy algorithms that increase the number of input variables or use them in real time.

8.4.10 Results and Conclusions

To demonstrate the use of the proposed hierarchical fuzzy system, a sample study is used to determine salary categories based on age, wit, and contact. The simulated sample is used to generate a total of 200 data points and serves as a set of input–output pairs to construct the rule base. The proposed algorithm is used to construct a hierarchical fuzzy rule base. After the first step of converting all numeric values into fuzzy rules, I used the membership function to generate the pursuing fuzzy rules:

Age (Age) = (Young (Y), Middle (M), Old (0));

Experience (Exp) = {Little (L), Some (S), Good (G)};

Contacts (Con) = {Poor (P), Normal (N), Quality (Q)};

Salary (Sal) = {Basic (B), Pearly (F), Upper (H))

R_1: If Age is Y and Exp is L and Con is P, then Sal is B

R_2: If Age is Y and Exp is L and Con is N, then Sal is B

R_3: If Age is Y and Exp is L and Con is Q, then Sal is B

R_4: If Age is Y and Exp is S and Con is P, then Sal is B

R_5: If Age is Y and Exp is S and Con is N, then Sal is B

R_6: If Age is Y and Exp is S and Con is Q then Sal is F

R_7: If Age is Y and Exp is G and Con is P, then Sal is H

R_8: If Age is Y and Exp is G and Con is N, then Sal is H

R_9: If Age is Y and Exp is G and Con is Q, then Sal is H

R_{10}: If Age is M and Exp is L and Con is P, then Sal is B

R_{11}: If Age is M and Exp is L and Con is N, then Sal is B

R_{12}: If Age is M and Exp is L and Con is Q, then Sal is F

R_{13}: If Age is M and Exp is S and Con is P, then Sal is F

R_{14}: If Age is M and Exp is S and Con is N, then Sal is F

R_{15}: If Age is M and Exp is S and Con is Q, then Sal is F

R_{16}: If Age is M and Exp is G and Con is P, then Sal is H

R_{17}: If Age is M and Exp is G and Con is N, then Sal is H

R_{18}: If Age is M and Exp is G and Con is Q, then Sal is H

R_{19}: If Age is 0 and Exp is L and Con is P, then Sal is B

R_{20}: If Age is 0 and Exp is L and Con is N, then Sal is F

R_{21}: If Age is 0 and Exp is L and Con is Q, then Sal is H

R_{22}: If Age is 0 and Exp is S and Con is P, then Sal is F

R_{23}: If Age is 0 and Exp is S and Con is N, then Sal is F

R_{24}: If Age is 0 and Exp is S and Con is Q, then Sal is H
R_{25}: If Age is 0 and Exp is G and Con is P, then Sal is H
R_{26}: If Age is 0 and Exp is G and Con is N, then Sal is H
R_{27}: If Age is 0 and Exp is G and Con is Q, then Sal is H

Regularity Yardstick measure is now performed on the dataset to determine the level of each input in the hierarchy. The Relative Comparison (RC) values of the three input variables are Age (338.584), Wits (149.102), and Contacts (495.492). The wattle of the level of hierarchy is Wits at the top, followed by Age and then Contacts. Without the hierarchy has been established, pruning is performed based on the two factors proposed earlier. The pursuit page shows the final hierarchical fuzzy rule wiring without pruning. Effectively, the total number of fuzzy rules in this hierarchical fuzzy rule wiring is 15, which is a notable reduction from the original 27 fuzzy rules.

R_1:

    If Exp is L, then use R_{21}
    If Exp is S, then use R_{22}
    If Exp is G, then Sal is H
R_{21}:

    If Age is Y, then Sal is B
    If Age is M, then use R_{32}
    If Age is 0, then use R_{33}
R_{22}:

    If Age is Y, then use R_{32}
    If Age is M, then Sal is F
    If Age is 0, then use R_{36}
R_{32}:

    If Con is Q, then Sal is F
    If Con is NOT Q, then Sal is B
R_{33}:

    If Con is P, then Sal is B
    If Con is Q, then Sal is H

R_{36}:

If Con is Q, then Sal is H
If Con is NOT Q, then Sal is F

8.5 Equalization of Time-Varying Nonlinear Digital Contacts

Digital cellular radio (DCR) contact systems face co-channel interference (CCI), proximal water supply interference (ACI), and inter-symbol interference (ISI) in the presence of additive white Gaussian noise (AWGN). CCI occurs due to frequency reuse, and the frequency spacing of various cells contributes to the ACI. These effects particularly confuse DCR, but the effects of ISI due to the narrow garland water supply nature generally confuse all digital contact systems. Adaptive equalizers are used to mitigate one or more of these effects on connected receivers. In DCR applications, CCI limits the performance of the equalizer. Linear fractional spacing equalizers (FSE) were often used for the equalization of these channels. This equalizer treats CCI as cyclic stop interference.

The performance of these equalizers is limited by the linear visualization Purlieus provided by the linear equalizer. Nonlinear technology was used for equalization of contact channels damaged by CCI and AWGN. Nonlinear equalizers offer superior performance compared to linear equalizers because they are worth forming nonlinear visualization boundaries. The high performance of these equalizers is not shaken by the Bayesian equalizer. Fuzzy filters are nonlinear filters and have been used for equalization in various connected systems.

There is a tropical relationship between recent Bayesian equalizers and fuzzy equalizers. Several nonlinear techniques for equalization of channels using CCI are misogynists. The performance of these nonlinear equalizer ids is limited to the low interference of the upper noise or the low noise of the upper interference (Chen et al. 1993). To mitigate the effects of CCI, ISI, and noise, we proposed a Bayesian decision feedback equalizer (DFE). Although this equalizer provides good performance, it has a lot of computational requirements.

8.5.1 Preparation for Equalization of Water Supply

The scalar constant and the scalar co-channel state of the equalizer can be unscientific by the clustering algorithm.

- Channel state: The equalizer vector constant state may be unscientific in the scalar constant state, and during the training period may not be shaken in the noisy receiving scalar using the supervised clustering algorithm. The noise is caused by the middle point of zero and the co-channel state as a positive and negative pair. The effect is canceled during the process of estimating the condition of the water supply. Signal-to-interference-noise-ratio (SINR) may be unscientific in supervised clustering during scalar water supply condition estimation. The estimated scalar tap water state can be retrieved along with the training signal sequence that produces it to form a vector tap water state.

• Common channel condition: When the water supply condition stops shaking, the water supply residue can be estimated. Water residues arise from CCI and AWGN. Unsupervised clustering algorithms such as k-means clustering or improved k-means clustering algorithms can be used to estimate scalar co-channel state and noise variance.

8.5.2 Why Type 2 FAFs Are Needed?

Radial basis function (RBF)–based neural networks have been successfully used to solve many nonlinear problems, including adaptive water supply equalization problems. There are a variety of adaptive fuzzy/neural fuzzy constant equalizers that closely fit the wholesale framework of RBF neural network–based systems. We often consider a type 2 fuzzy adaptive filter (FAF)–based water equalizer with compensatory neuro-fuzzy filter (CNFF) based on a water supply equalizer and one based on the adaptive network-based fuzzy inference system (ANIFS) as unromantic for mobile cellular channels. It can be seen that the implementation of the adaptive equalizer fits into the general framework of an RBF-based system.

8.5.3 FAF Design

Functional equivalence between RBF neural networks and a simplified admission of fuzzy inference systems has been created, which allows us to use what we find for one of the models (learning rules, expressive powers, etc.) for another model. It is interesting to observe that the two models originating from various origins are functionally equivalent. Although these two models were motivated from various origins (RBF networks in physiology and fuzzy inference systems in cognitive reasoning systems), they share global characteristics not only in working with data but also in the learning process of decompressing the desired mapping. We show that they are functionally equivalent under some minor restrictions. Learning algorithms and theorem for the expressiveness of one model can be unromantic to another model and vice versa. The output of the RBF network can be calculated in two ways.

If pursuit is true, you can establish functional equivalence between the RBF network and the fuzzy inference system.

- The number of accepted field units is equal to the number of fuzzy IF–THEN rules.
- The output of each fuzzy IF–THEN rule is a well-balanced constant.
- Membership functions within each rule are chosen as Gaussian functions with the same variance.
- The t-norm operator used to calculate the strength of each rule execution is multiplication (or product t-norm).
- The RBF neural network and fuzzy inference system under consideration both use the same method (i.e. weighted stereotype or weighted sum) to derive the overall output.

By now, if you can build an integrated framework for adaptive equalizers based on neural networks and fuzzy logic, you will have some convenience. Combining the two heavier features will have a synergistic effect on improving performance. We will not cover the three adaptive equalizers we are considering one by one. Traditional adaptive algorithms for equalizers are based on a measure that minimizes the midpoint squared error between the desired filter output and the immediate filter output (Chen et al., 1993). We investigated the use of a radial base function network for equalization of digital communications water supply.

It shows that the RBF network has the same structure as the optimal Bayesian symbolic decision equalizer solution and can therefore be used to implement a Bayesian equalizer. Training of the RBF network to realize Bayesian equalization solution can be achieved efficiently using a simple and powerful supervised clustering algorithm. This represents a radically new flow of adaptive equalizer design.

8.5.4 Simulation and Conclusion

The most effective way to design a fuzzy intelligent system assumes a generalized strategy, which is a combination of bottom-up and top-down design approaches. For example, the same as the first assistant (bottom-up design), we separate the tracery of the fuzzy intelligent system (fuzzy ES or fuzzy controller) supported by the power software and hardware realization. The same as the second assistant (top-down design), it separates the algebra of the T-norm and T-conorm-based operator families, one of the most unqualified standard knowledge algebras in the symbolic system. Then, use the T-norm parameter implicit function to generate a T-norm-based inference procedure. Here, the language of fuzzy scale is used to represent the uncertainty of the fuzzy knowledge base. These generalized fuzzy inference algorithms provide an increasingly undefined methodology to the diamond of fuzzy logic controllers. Similar non-standard fuzzy inference algorithms can be created for fuzzy models based on fuzzy models and fuzzy non-monotonic approaches.

The quality of the obtained logic can be investigated by means of simulation. To this end, the Mamdani fuzzy inference system (SMFI) simulation module is designed to simulate changes in the behavior of a fuzzy expert system (ES) when increasing the logic in the inference module.

While designing the ES, previous systems were unfamiliar with the use of prognosis for inference results. A typical ES requires minimal knowledge wiring to demonstrate its accuracy. In most systems, fuzzy inference methods are still in stock at design time and usually do not change while using the system.

The main problem with fuzzy reasoning systems is linguistic trait errors. The results of each reasoning step generally do not match one of the language scale factors, so we use a linguistic approximation. In multiple steps of inference, these approximations accumulate and the error in the result became 1 (impossible) or 0 (floats). The number of these steps depends on the knowledge wiring and fuzzy logic. The idea of our model is to separate the logic to infer as long as possible.

SMFI offers the possibility to create false prognosis for inference results using multiple parameters of a given language scale, fuzzy logic, and knowledge base. As the main model for the simulation, the fuzzy problem reduction method given by AND/OR graph was chosen. Assume that every node has only two inputs. Get a prognosis. It separates random nominations from this kind of set of trees (1000 or less) and then evaluates the median of the fuzzy values of the inference results. The input information is the median of the linguistic meaning of the facts in the knowledge connection and the median of the linguistic meaning of the rule.

- Dependence on the knowledge base
- Dependence on the depth of the inference tree
- Dependence on the shape of the inference tree
- Dependence on T-norm

8.6 Liaison System with ISI and CCI

The performance of two multichannel receivers, namely the FTS-MR and the TWN-MLSE-MR – recently ripened and patented by Thomson-CSF-Communications – can be evaluated and compared in the presence of IS1 and CCI, for variegated scenario of useful signal and interferences. The two receivers are very powerful, but despite of the fact that the TWN-MLSE-MR is asymptotically larger than the FTS-MR in all cases, the FTS-MR is increasingly powerful than the other MR in a TU environment when a strong interference is present. This result is due to the total noise correlation matrix estimation, required for the implementation of the TWN-MLSE-MR, which is far from the stuff perfect for a strong Intelligence Surveillance Reconnaisance (ISR).

8.7 Connection Ticket Rental for ATM Networks

Today's communication technology goes through a hair-trigger curve. Since the introduction of the ISDN business, a lot of standards have been defined for the broadband network. It was well established at the time that the future broadband would be ATM transfer mode for ISDN. As reliability is sufficient with the need for a flexible network and the advancement of technology and computer concepts, ISDN's ultimate solution is to use ATMs. At higher speeds, ATM networks can handle traffic for diversity and quality of service (QoS) requirements.

Spirits (traffic and its connections) between O-D in ATM environments are a direction and only towards the original first destination. Connection Ticket Leasing in ATM networks depends on a set of factors such as the characteristics of the traffic and the network resources. Traffic characteristics in the traditional Connection Ticket Leasing must be verified, and the required network resources (i.e. bandwidth) must be verified for any violation of the judgment of the ticket. It seems that this process is not identical with the ATM environment, which is aimed at stopping all re-transactions and concentrating spirits in the same direction. CAC results are constructed based on the user's notifications of his or her traffic such as the lamina rate of the stereotype and the splash length for data communications. In many cases, these announcements are inaccurate, resulting in a waste or lack of resources, and thus leading to data corruption. To free the network from user misrepresentation, it was surprising that CAC measured the number of incoming cells to help it make accurate decisions. But unfortunately this policy only applies to peak lamina rate declaration. In accordance with the shortcomings mentioned earlier, the time spent checking the network resources is sufficient to handle the waiting arrival.

To prevent mentioned deficiencies, we propose a new CAC that combines both user and calibration notification for traffic specifications. The proposed strategy is to customize the traffic specifications without having to access the virtual connection, which is embedded in the node (sub-memory). Therefore, without the traffic being unsafe, it will be checked and prepared for transfer via physical connection.

8.7.1 Survey-Based CAC Using Type 2 FLS: Overview

A well-established (QoS) maintenance of the connection once installed on the network and the availability of sufficient bandwidth to handle the waiting undeniable quality of its requested quality (QoS) must wait until the CAC verifies that a new arrival is still undeniable. These trommels or trial hoops of routing tables for all outgoing connections at the source node and the next node indicate a lot of time consumption and loads for the processing unit. In the more weighted model, if the result of the trammels is positive, we have to pay the flipside time to determine the surprising denial in its virtual channel (VC). All of these processing functions slow down the exchange of information. Link total water supply topics are undoubtedly the main limit on incoming calls. If the bandwidth check is not sent, you will have to wait at the entrance of the denial terminal. Can we use the time spent verifying the availability of resources and causing denials to be sent via the misogynist link?

One can aim to cross the line of link total water supply topics in a short time. Instead of waiting at the entrance of the terminal, it is too much to use the waiting time (which is wasted) in preparing the undeniable for the abduction. This idea can be realized by setting up a subsidiary storage company to represent the physical connection, so the physical connection is replaced by a virtual connection, but with very high efficiency. This virtual connection will be separated just like the physical connection. The virtual path identifiers (VPIs) are identical except that the bandwidth of the virtual path (VP) of the virtual connection is higher than that of the physical connection. The bandwidth (size) of the VC will be the same for both connections, but their size will be different (we use the route and service separation).

As a result, we created a physical connection-like shadow connection with the same number of VPs but with larger bands and a larger number of VCs of the same bandwidth. The proposed virtual connection is in the subconscious mind at the tip of a sub-memory. Virtual connection is divided into sections. They are each recommended to serve a specific matriculation, so the number of these sections equals the number of transport classes. The virtual connection (sub-memory) is divided into these sections.

8.7.2 Extraction of Knowledge for CAC

Asynchronous transmission system is a high-speed switching technology in broadband communication networks (P-ISDNs). There are many types of communication services such as audio, video, or high-speed data that can be carried through these ATM networks. With its ability to classify unsorted bit rate (rt-VBR) or real-time (nrt-VBR) constant bit rate, variable bit rate classified services. ATM technology is rapidly being recognized as a kind of communication barometer of the future, so high-speed ATM services can be achieved everywhere.

Connection Ticket Leasing in ATM networks is a spirits executive function that decides whether to engage or determine a new connection to engage in the network. This visualization is usually based on the network load of the current ATM network, idle network resources (output bandwidth, buffer zone size, etc.), spirits parameters, and the quality of the registered service of new connections on the network and on existing connections. Definitions of spirits parameters in ATM networks include peak lamina rate, fixed lamina rate, maximum splash size, and minimum cell rate (MCR). Include. To ensure the quality of service, an "agreement" must be signed between the user and the network service provider when a connection is established, which clearly defines the spirits parameters and the quality of the service. In addition, other parameters related to service quality are specified in ATM networks, such as Lamina Loss Rate, Lamina Weight Variants, and Maximum Lamina Transfer Weight.

Several models of CAC have been proposed so far. These CAC models can be classified into two types: CAC based on measurement and CAC based on spirits interpretation. The latter uses features provided by sources during the setup phase, summarizing the bad model for existing links and newly allowed connections. Therefore, the implementation of the CAC based on the spirits description is based on the verification of the parameters provided by the source when a connection is established. However, it is difficult for users to describe what parameter or maximum splash length parameter is the stereotype pocket speed. In addition, to provide a well-defined QoS range, the CAC based on the spirits interpretation must penetrate the bandwidth for the worst cases. Therefore, using this CAC model is successful when the spirits are soft; however, when the spirits explode, the ability to use the bandwidth is limited.

8.7.3 Survey Process

Graphics processing units (GPUs) have been advancing rapidly over the past decade. The hardware used first for the image processing displayed on the screen has been converted into a device for calculating parallel (general-purpose GPU). The GPU can be used to perform radar data processing during the signal processing phase or during the data processing stage. Considering that radar data are processed on a large scale, it is washed away and allows the computational process to be parallel. Previously, radar data processing was done using a special digital signal processor (DSP) device and/or field programmable gate array (FGGA). But the confiscation required for both types of devices is more expensive than the GPU. In addition, both have low scaling.

Today, high-performance embedded computer applications pose significant challenges. Traditionally, the application has been realized in the use of application-specific integrated circuit (ASIC) and/or DSPs. The advantages of ASIC in terms of performance and power requirements can often not keep up with the increase in fiction costs. GPUs, on the other hand, offer superior computer performance while maintaining their versatility.

Radar is a system that detects the presence, direction, distance, and speed of aircraft, ships, and other objects (moving and stationary objects) by sending radio waves that reflect from object to source. Radar transmits a short pulse modified by a directional antenna, which rotates at an uncontrolled speed. If the antenna leads to the target, a small portion of the transmitted energy will be reflected from the target as it travels to the antenna. The waiting distance between sent and received legumes is proportional to the distance. The direction of the antenna (azimuth antenna) determines the azimuth of the target in the process of placing the gingerbread. Image produced by radar helps in locating an object that reflects the signal. Traditionally, source images are provided with a special CRT view tabbed project-level indicator. Currently, source images are provided using computers and raster displays.

In high-frequency (HF) radar systems, the classical platform for signal processing uses DSP and field programmable gate array (FPGA). However, the seizure of these two sites is very high, and the scalability is bad for both sites. It is therefore important to find new ways to process HF radar signals. Inspired by the software-limited radar concept and the excellent hardware speed performed in parallel processing to the GPU, we explored how to implement the HF radar algorithm on a GPU on a regular PC platform. Numerical tests were carried out to demonstrate the effectiveness of the proposed method.

8.7.4 CAC Visualization Boundaries and Results

Due to the need for all types of communication services such as video, voice, and data, the automated teller machine (ATM) network has gained top attention. The health of ATM networks is the flexibility to control different types of traffic with different transport characteristics and QoS requirements. A set of traffic leasing functions must be provided by the ATM network to confirm the QoS of each service and unpack the upper network usage.

One of the transport leasing functions for the ATM system is the connection admission control (CAC). The CAC is referred to as "the set of deportations taken by the network during the undeniable setup phase to determine whether the connection request can be surprised or rejected." When a new connection is requested, the ticket controller examines its QoS requirements, traffic characteristics parameters, and the current network status to determine whether the new connection should win.

Several CAC strategies have been proposed for ATM networks. Kukrin et al. proposed the equivalent heading method to estimate the bandwidth required for individual or block connections under the control of a given QoS requirement. Cyto proposed an undeniable ticket scheme by assuming the lamina over the probability of loss from the traffic parameters specified by the users. However, the rating error will accumulate in these projects provided a simple bandwidth working principle by classifying all transport sources. However, it would not be wise to have a new source of traffic that does not belong to any predefined class of transport.

Ambiguous logic systems are widely used to deal with CAC-related problems in ATM networks. They provide a robust mathematical framework for dealing with real-world inaccuracies, and walk out soft principles that are highly qualified to adapt to dynamic, inaccurate, and explosive environments. Bongle and Ghosh used ambiguous mathematics to provide a flexible, high-performance solution for queue management in ATM networks (Chang et al., 1997). An ambiguous traffic controller that simultaneously includes CAC and congestion control was proposed. Comparative studies show that the proposed ambiguous approaches significantly improve system performance compared to conventional approaches. Furthermore, the self-learning skills of neural networks are exceptional to deal with CAC-related problems. Hiramatsu used a neural network as a link ticket controller. Tran-Kia and Grop explored the use of a neural network to perform CAC. Simulation results showed that neural networks are feasible in implementing ATM traffic control.

Most of the proposed CAC methods are based on time-domain analysis. Li et al. studied the sequence performance of an ATM network using the power-spectrum parameters (i.e. frequency-domain characteristics) of input traffic (Chong et al., 1995). The power spectrum can capture the contact and burst aspects of the input process in the time domain. The low-frequency evening sequence of the input power-spectrum dominates the performance, and the effect of the HF evening spectrum can be ignored. As the low-frequency component increases, so does the source of traffic. Information on input traffic in the frequency domain can be easily retrieved using DSP chips. Coming from the power-spectrum perspective of transportation may be an optional recommendation, and a power-spectrum-based CAC system would be constructive for the ATM network. The power-spectrum-based CAC method, which uses power-spectrum parameters as symbols, was proposed to create the CAC view table. However, this required a large amount of memory and was not possible in some ATM systems. The power-spectrum-based neural-net CAC (PNCAC) method was proposed to create the CAC's optimal visualization hyperplane to modify the vision table. The disadvantage is that the knowledge categorized in the previously proposed methods is difficult to integrate into the diamond of a neural network.

8.8 Exercise

1. Write a short note on rule-based nomenclature of video traffic.
2. What do you imply by CCI and ISI? How can they be overcome?
3. Discuss fuzzy rules. Explain by an example.
4. What is simulation? Why is it important in fuzzy logic system?
5. Discuss waterworks equalization.

References

Aladi, Jabran Hussain, Christian Wagner, and Jonathan M. Garibaldi. "Type-1 or interval type-2 fuzzy logic systems: On the relationship of the amount of uncertainty and FOU size." *In 2014 IEEE International Conference on Fuzzy Systems (FUZZ-IEEE)*, IEEE, Beijing, China (2014), pp. 2360–2367.

Chang, Chung-Ju, Song-Yoar Lin, Yow-Ren Shiue, and Ray-Guang Cheng. "A power-spectrum based neural fuzzy connection admission mechanism for ATM networks." *In Proceedings of ICC'97-International Conference on Communications*, IEEE, Montreal, Quebec, Canada (1997), vol. 3, pp. 1709–1713.

Chen, Sheng, Bernard Mulgrew, and Peter M. Grant. "A clustering technique for digital communications channel equalization using radial basis function networks." *IEEE Transactions on Neural Networks* 4, no. 4 (1993): 570–590.

Chong, Song, San-qi Li, and Joydeep Ghosh. "Predictive dynamic bandwidth allocation for efficient transport of real-time VBR video over ATM." *IEEE Journal on Selected Areas in Communications* 13, no. 1 (1995): 12–23.

Kwakernaak, Huibert. "Fuzzy random variables: I. Definitions and theorems." *Information Sciences* 15, no. 1 (1978): 1–29.

Mendel, Jerry M., and Feilong Liu. "Super-exponential convergence of the Karnik–Mendel algorithms for computing the centroid of an interval type-2 fuzzy set." *IEEE Transactions on Fuzzy Systems* 15, no. 2 (2007): 309–320.

Nie, Maowen, and Woei Wan Tan. "Towards an efficient type-reduction method for interval type-2 fuzzy logic systems." *In 2008 IEEE International Conference on Fuzzy Systems (IEEE World Congress on Computational Intelligence)*, IEEE, Hong Kong (2008), pp. 1425–1432.

Shotton, Jamie, Ben Glocker, Christopher Zach, Shahram Izadi, Antonio Criminisi, and Andrew Fitzgibbon. "Scene coordinate regression forests for camera relocalization in RGB-D images." *In Proceedings of the IEEE Conference on Computer Vision and Pattern Recognition*, Washington, DC (2013), pp. 2930–2937.

Appendix A: Join, Meet, and Negation Operations for Non-Interval Type 2 Fuzzy Sets

A.1 Introduction

General type 2 fuzzy sets (GT2FSs) are characterized by secondary memberships, which take any value between 0 and 1 (unlike interval type 2 fuzzy sets (IT2FSs), whose secondary memberships are either 0 or 1). The meet and join operations for GT2FSs, which represent the intersection and union for these sets, respectively, are supported by the extension principle by Zadeh as generalization of the intersection and union for type 1 fuzzy sets. In 2001, the initial work by Karnik and Mendel presented a simplified procedure to compute these operations for GT2FSs, although it trusted the condition that the secondary grades of type 2 fuzzy sets were normal and convex type 1 fuzzy sets. This work was later generalized by Coupland and John to include non-normal sets by borrowing some methods (Weiler–Atherton, modified Weiler Atherton, Bentley–Ottmann Plane Sweep Algorithm, etc.) from the sector of computational geometry; yet, convexity remained a necessary condition. Newer works studied the geometrical properties of some GT2FSs to seek out closed formulas or approximations for the join and meet operations in some specific cases.

Recent developments in type 2 symbolic logic have changed the perception researchers have of IT2FSs. IT2FSs are type 2 fuzzy sets in which uncertainty is equally distributed within the dimension (also called secondary membership), and thus, these secondary memberships are either 0 or 1, unlike GT2FSs, in which uncertainty within the dimension isn't equally weighted and, therefore, the distribution is often an arbitrary type 1 fuzzy set. When IT2FSs were initially defined, all the idea and operations were supported the precise case where IT2FSs are like interval-valued fuzzy sets (IVFSs). However, it's been recently shown that IT2FSs are more general than IVFSs. Hence, so as to derive the idea of those general sorts of interval type 2 symbolic logic systems (IT2FLSs) (which employ IT2FSs which aren't like IVFSs), it's necessary to develop the meet and join operations of GT2FSs with non-convex secondary memberships and, then, particularize it to the case of IT2FSs, which have secondary grades adequate to either 0 or 1. Hence, we'll be finding the join and meet operations for

GT2FSs where secondary memberships are arbitrary type 1 sets and, hence, are often non-convex and/or non-normal. This may be wont to derive the join and meet operations of IT2FSs where the secondary grades are non-convex sets.

T-norm fuzzy logics are a family of non-classical logics, informally delimited by having a semantics that takes the important unit interval [0, 1] for the system of truth values and functions called t-norms for permissible interpretations of conjunction. They're mainly utilized in applied symbolic logic and fuzzy pure mathematics as a theoretical basis for approximate reasoning.

T-norm fuzzy logics belong in broader classes of fuzzy logics and many-valued logics. So as to get a well-behaved implication, the t-norms are usually required to be left-continuous; logics of left-continuous t-norms further belong within the class of sub-structural logics, among which they're marked with the validity of the law of prelinearity, $(A \rightarrow B) \vee (B \rightarrow A)$. Both propositional and first-order (or higher-order) t-norm fuzzy logics, as well as their expansions by modal and other operators, are studied. Logics that restrict the t-norm semantics to a subset of the important unit intervals (for example, finitely valued Łukasiewicz logics) are also usually included within the class.

Important samples of t-norm fuzzy logics are monoidal t-norm logic MTL of all left-continuous t-norms, basic logic BL of all continuous t-norms, product symbolic logic of the merchandise t-norm, or the nilpotent minimum logic of the nilpotent minimum t-norm. Some independently motivated logics belong among t-norm fuzzy logics, too; for instance, Łukasiewicz logic (which is that the logic of the Łukasiewicz t-norm) or Gödel–Dummett logic (which is that the logic of the minimum t-norm).

In a partially ordered set P, the join and meet of a subset S are, respectively, the supremum (least upper bound) of S, denoted as ∨S, and infimum (greatest lower bound) of S, denoted as ∧S. Generally, the join and meet of a subset of a partially ordered set needn't exist; once they do exist, they're elements of P.

Join and meet can also be defined as a commutative, associative, and idempotent partial Boolean operation on pairs of elements from P. If a and b are elements from P, the join is denoted as a ∨ b and, therefore, the meet is denoted as a ∧ b.

Join and meet are symmetric duals with reference to order inversion. The join/meet of a subset of a completely ordered set is just its maximal/minimal element, if such a component exists.

A partially ordered set during which all pairs have a join may be a join-semilattice. Dually, a partially ordered set during which all pairs have a meet may be a meet-semilattice. A partially ordered set that's both a join-semilattice and a meet-semilattice may be a lattice. A lattice during which every subset, not just every pair, possesses a meet and a join may be a complete lattice. It's also possible to define a partial lattice, during which not all pairs have a meet or a join, but the operations (when defined) satisfy certain axioms.

A.2 Join

Join may be a lattice-theoretic concept that requires not to have anything to try to with unions. As an example, the positive integers partially ordered by divisibility are a lattice during which the join of two integers is their least integer. An example is R^2 partially ordered in order that

$$(a,b) \le (c,d) \text{ if } a \le c \text{ and } b \le d;$$

in that lattice

$$(a,b) \vee (c,d) = (\max\{a,c\}, \max\{b,d\}),$$

If X may be a set, and T is that the set of all topologies on X, $\langle T, \subseteq \rangle$ may be a lattice, the join of two topologies generally isn't their union: rather, it's the topology generated by taking their union as a sub-base.

A.3 Meet

Meet may be a lattice-theoretic concept that requires not to have anything to try to with intersection. As an example, the positive integers partially ordered by divisibility are a lattice during which the meet of two integers is their greatest common factor. An example is R^2 partially ordered in order that

$$(a,b) \le (c,d) \text{ if } a \le c \text{ and } b \le d;$$

in that lattice

$$(a,b) \wedge (c,d) = (\min\{a,c\}, \min\{b,d\}).$$

A.4 Negation

The membership function of the complement of a fuzzy set A with membership function μA is defined as the negation of the specified membership function. This is called the negation criterion. The Complement operation in fuzzy set theory is the equivalent of the NOT operation in Boolean algebra.

Appendix B: Properties of Type 1 and Type 2 Fuzzy Sets

B.1 Introduction

Fuzzy set enables one to work in uncertain and ambiguous situations and solve ill-posed problems or problems with incomplete information.

Human visual system is perfectly adapted to handle uncertain information in both data and knowledge. It will be hard to define quantitatively how an object, such as a car, has to look in terms of geometrical primitives with exact shapes, dimensions, and colors. We use descriptive language to define features that eventually are subject to a wide range of variations.

Let us consider fuzzy set "two or so." In this instance, universal set X are the positive real numbers: $X = \{1, 2, 3, 4, 5, 6, \ldots\}$. Membership function for A = "two or so" in this universal set X is given as follows: $\mu A(1) = 0.5$, $\mu A(2) = 1$, $\mu A(3) = 0.5$, $\mu A(4) = 0$.

Fuzzy set theory was formalized by Professor Lofti Zadeh at the University of California in 1965. What Zadeh proposed is very much a paradigm shift that first gained acceptance in the Far East, and its successful application has ensured its adoption around the world.

A paradigm is a set of rules and regulations that define boundaries and tell us what to do to be successful in solving problems within these boundaries. For example, the use of transistors instead of vacuum tubes is a paradigm shift; likewise, the development of fuzzy set theory from conventional bivalent set theory is a paradigm shift.

B.2 Type 1 Fuzzy Sets

In type 1 fuzzy sets, each element is mapped to [0, 1] by a membership function: $\mu A: X \rightarrow [0, 1]$, where [0, 1] means real numbers between 0 and 1 (including 0 and 1).

In a type 1 fuzzy logic system (FLS), the inference engine combines rules and gives a mapping from input type 1 fuzzy sets to output type 1 fuzzy sets. Multiple antecedents in rules are connected by the t-norm (corresponding to intersection of sets). The membership grades in the input sets are combined with those in the output sets using the sup-star composition. Multiple rules may be combined using the t-conorm operation (corresponding to union of sets) or during defuzzification by weighted summation. In the type 2 case, the inference process is very similar. The inference engine combines rules and gives a mapping from input type 2 fuzzy sets to output type 2 fuzzy sets. To do this, one needs to find unions and intersections of type 2 sets, as well as compositions of type 2 relations. In a type 1 FLS, the defuzzifier produces a crisp output from the fuzzy set that is the output of the inference engine, i.e., a type 0 (crisp) output is obtained from a type 1 set. In the type 2 case, the output of the inference engine is a type 2 set; so, we use extended versions (using Zadeh's extension principle of type 1 defuzzification methods). This extended defuzzification gives a type 1 fuzzy set. Since this operation takes us from the type 2 output sets of the FLS to a type 1 set, we call this operation type reduction, and so, the type-reduced set obtained a type-reduced set.

B.3 Type 2 Fuzzy Sets

Type 2 fuzzy sets and systems generalize standard type 1 fuzzy sets and systems so that more uncertainty can be handled. From the very beginning of fuzzy sets, criticism was made about the fact that the membership function of a type 1 fuzzy set has no uncertainty associated with it, something that seems to contradict the word fuzzy, since that word has the connotation of lots of uncertainty. So, what does one do when there is uncertainty about the value of the membership function? The answer to this question was provided in 1975 by the inventor of fuzzy sets, Prof. Lotfi A. Zadeh, when he proposed more sophisticated kinds of fuzzy sets, the first of which he called a type 2 fuzzy set. A type 2 fuzzy set lets us incorporate uncertainty about the membership function into fuzzy set theory, and is a way to address the above criticism of type 1 fuzzy sets head-on. And, if there is no uncertainty, then a type 2 fuzzy set reduces to a type 1 fuzzy set, which is analogous to probability reducing to determinism when unpredictability vanishes.

In order to symbolically distinguish between a type 1 fuzzy set and a type 2 fuzzy set, a tilde symbol is put over the symbol for the fuzzy set; so, A denotes a type 1 fuzzy set, whereas $\tilde{A}$ denotes the comparable type 2 fuzzy set. When the latter is done, the resulting type 2 fuzzy set is called a general type 2 fuzzy set (to distinguish it from the special interval type 2 fuzzy set).

Prof. Zadeh didn't stop with type 2 fuzzy sets, because in that 1976 paper, he also generalized all of this to type n fuzzy sets. The membership function of a general type 2 fuzzy set, $\tilde{A}$, is three-dimensional, where the third dimension is the value of the membership function at each point on its two-dimensional domain that is called its footprint of uncertainty (FOU).

For an interval type 2 fuzzy set, that third-dimension value is the same (e.g., 1) everywhere, which means that no new information is contained in the third dimension of an interval type 2 fuzzy set. So, for such a set, the third dimension is ignored, and only the FOU is used to describe it. It is for this reason that an interval type 2 fuzzy set is sometimes called a first-order uncertainty fuzzy set model, whereas a general type 2 fuzzy set (with its useful third dimension) is sometimes referred to as a second-order uncertainty fuzzy set model.

The FOU represents the blurring of a type 1 membership function, and it is completely described by its two bounding functions: a lower membership function (LMF) and an upper membership function (UMF), both of which are type 1 fuzzy sets. Consequently, it is possible to use type 1 fuzzy set mathematics to characterize and work with interval type 2 fuzzy sets. This means that engineers and scientists who already know type 1 fuzzy sets will not have to invest a lot of time learning about general type 2 fuzzy set mathematics in order to understand and use interval type 2 fuzzy sets.

Work on type 2 fuzzy sets languished in the period from the 1980s to the mid-1990s, although a small number of articles were published about them. People were still trying to figure out what to do with type 1 fuzzy sets, so even though Zadeh proposed type 2 fuzzy sets in 1976, the time was not right for researchers to drop what they were doing with type 1 fuzzy sets to focus on type 2 fuzzy sets. This changed in the latter part of the 1990s as a result of Prof. Jerry Mendel and his students' works on type 2 fuzzy sets and systems. Since then, more and more researchers around the world are writing articles about type 2 fuzzy sets and systems.

Appendix C: Computation

The problem is to estimate the extent of risk involved during a software engineering project. For the sake of simplicity, we'll reach our conclusion supported by two inputs: project funding and project staffing.

C.1 Introduction

The first step is to convert the crisp input into a fuzzy one. Since we've two inputs, we'll have two crisp values to convert. The first value is the level of project staffing. The second value is the level of project funding. Suppose our inputs are project_funding = 35% and project_staffing = 60%. We can get the fuzzy values for these crisp values by using the membership functions of the acceptable sets. The sets defined for project_funding are inadequate, marginal, and adequate. The sets defined for project_staffing are small and enormous.

Thus, we've the following fuzzy values for project_funding:

$$\mu_{\text{funding = inadequate}}(35) = 0.5$$

$$\mu_{\text{funding = marginal}}(35) = 0.2$$

$$\mu_{\text{funding = adequate}}(35) = 0.0$$

C.2 The Rules

Now that we've the fuzzy values, we will use the fuzzy rules to reach the ultimate fuzzy value. The rules are as follows:

1. If project_funding is adequate or project_staffing is little, then risk is low.
2. If project_funding is marginal and project_staffing is large, then risk is normal.
3. If project_funding is insufficient, then risk is high.

Rule 1: If project_funding is adequate or project_staffing is small, then risk is low

Rules containing disjunctions, OR, are evaluated using the UNION operator.

$$\mu_{A \cup B}(x) = \max[\mu_A(x), \mu_B(x)]$$

$$\mu_{\text{risk = low}} = \max[\mu_{\text{funding = adequate}}(35), \mu_{\text{staffing = small}}(60)] = \max[0.0, 0.1] = 0.1$$

Rule 2: If project_funding is marginal and project_staffing is large, then risk is normal.

Conjunctions in fuzzy rules are evaluated using the INTERSECTION operator.

$$\mu_{A \cap B}(x) = \min[\mu_A(x), \mu_B(x)]$$

$$\mu_{\text{risk = normal}} = \max[\mu_{\text{funding = marginal}}(35), \mu_{\text{staffing = large}}(60)] = \max[0.2, 0.7] = 0.2$$

Rule 3: If project_funding is insufficient, then risk is high:

$$\mu_{\text{risk = normal}} = 0.2 * 0.7 = 0.14$$

C.3 Rule Evaluation Results

The result of evaluating the rules is shown below:

$$\mu_{\text{risk = low}}(z) = 0.1$$

$$\mu_{\text{risk = normal}}(z) = 0.2$$

$$\mu_{\text{risk = high}}(z) = 0.5$$

We now use the results to scale or clip the resultant membership functions. Once again for the sake of simplicity, we'll clip each of the functions:

$$\mu_{\text{risk=low}}(z) = 0.1$$

We perform a union on all of the scaled functions to get the ultimate result.

C.4 Defuzzification

The defuzzification is often performed in several alternative ways. The most popular method is the centroid method.

C.5 Centroid method

This method calculates the middle of gravity for area under the curve:

$$COG = \frac{\sum_{x=a}^{b} \mu_A(\chi)x}{\sum_{x=a}^{b} \mu_A(\chi)}$$

Bisector
   A vertical line that divides the region into two sub-regions of equal area. It is sometimes, but not always, coincident with the centroid line.
Mean of maximum
   Assuming there's a plateau at the utmost value of the ultimate function take the mean of the values it spans.
Smallest value of maximum
   Assuming there's a plateau at the utmost value of the ultimate function take the littlest of the values it spans.
Largest value of maximum
   Assuming there's a plateau at the utmost value of the ultimate function take the most important of the values it spans.
   We chose the centroid method to seek out the ultimate non-fuzzy risk value related to our project. This is shown below:

$$COG = \frac{(0+10+20)*0.1 + (30+40+50+60)*0.2 + (70+80+90+100)*0.5}{0.1*3 + 0.2*4 + 0.5*4}$$

$$= 67.4$$

The result is that this project has 67.4% risk related to it given the definitions above.

Appendix D: Medical Diagnosis by Fuzzy Logic

D.1 Idea of Crisp Set and Fuzzy Set

A set that provides an accurate starting point for developing a conception about something is known as a fuzzy set.

And a crisp set is a set defined in such a way that it contains two classes: one contains the elements belonging to a particular set, and the other contains the elements not belonging to the given set.

Let us now understand this with the help of a real-life example. Suppose I take a box of biscuits. Now I want to eat all the biscuits from the box, without knowing how many biscuits are there. Now, this is an example of fuzzy set, as the number of biscuits in the box is undetermined.

Now again if I say that I want to have two biscuits from the box. Then, this becomes a crisp set as the biscuits which I am not eating are the members of the second class which contains the elements not belonging to the set (Table D.1).

D.1.1 Definition of Fuzzy Sets

Fuzzy set in mathematics is a kind of set that contains elements having a certain degree of membership.

A fuzzy set gives us the accuracy of a certain value when it is tending towards the most accurate one (or height of the fuzzy set). This could be properly understood with the implementation of membership function. For example,

TABLE D.1

Differences between Fuzzy Sets and Crisp Sets

Fuzzy Sets	Crisp Sets
Defined for indeterminate entities	Defined for fixed determinable entities
Follows the infinite-valued logic	Follows the bivalued logic
Elements are partly accommodated by the sets	Elements can have a total membership or non-membership

$$\left\{ \frac{0.5}{2} \frac{0.3}{6} \frac{1}{4} \frac{0.2}{8} \right\}$$

where 2 and 6 have memberships 0.5 and 0.3, respectively, whereas 4 has membership 1.

D.1.2 Membership Function

A function that deals with the accuracy of a number when it is tending towards the height of a fuzzy set is known as a membership function.

D.1.3 Different Types of Membership Function

D.1.3.1 Intervalued Fuzzy Sets

A fuzzy set where the membership grade of an element is not just a real number that belongs to [0, 1] but an interval that belongs to $P[0,1]$ where $P[0,1]$ denotes all possible partitions/intervals within [0, 1].

Where ordinary fuzzy sets are represented by $A : X \rightarrow [0,1]$; these kind of fuzzy sets are represented as $A : X \rightarrow \varepsilon[0,1]$, where $\varepsilon[0, 1] \subset P[0, 1]$.

This can only be done when we don't know the actual values of the membership grades but do know the upper and lower bounds where they can possibly belong.

D.1.3.2 Fuzzy Set of Type 2

These are interval-valued fuzzy sets, where the intervals assigned to the elements are also ordinary fuzzy sets.

D.1.3.3 Normal Fuzzy Sets

These are fuzzy sets whose height is equal to 1: i.e. $h(A) = 1$

D.1.3.4 Subnormal Fuzzy Sets

These are fuzzy sets where $h(A) < 1$ are known as subnormal fuzzy sets.

D.2 Intuitionistic Fuzzy Sets

The qualitative measurements in the field of mathematics can be done through fuzzy methods. The idea of these kinds of set was established by Lotfi A. Zadeh and Dieter Klaua in 1965 and extended to the classical notion of set.

Intuitionistic fuzzy sets (IFS) have been introduced by Krassimir Atanassov (1983) as an extension of Lotfi Zadeh's notion of fuzzy set. Many work-life decision-making and applications in different portions of sciences using IFS are observed in the literature. One such application is due to the interpretation of medical sciences.

D.2.1 Basic Concepts of Intuitionistic Fuzzy Sets

IFS are sets whose elements have a degree of membership or non-membership and even a degree of hesitation or uncertainty. Usually, it is an extension of normal fuzzy sets.

Let a set E be fixed. An IFS A in E is an object having the following form:

$$A = \left\{ \left(x, \mu_A(x), v_A(x) \right) : x \in E \right\}$$

where the functions $\mu_A : E \to [0,1]$ and $v_A : E \to [0,1]$ define the degree of membership and the degree of non-membership, respectively, of the element $x \in E$ to the set A, which is a subset of E and for every $x \in E$, $0 \le \mu_A(x) + v_A(x) \le 1$. The amount $\pi_A(x) = 1 - \left(\mu_A(x) + v_A(x) \right)$ is called the hesitation part, which may cater to either membership value or non-membership value or both.

D.2.2 Some Other Properties of IFS

If A and B are two IFSs of the set E, then

i. $A \subset B$ iff $\forall x \in E$, $\mu_A(x) \le \mu_B(x)$ and $v_A(x) \ge v_B(x)$

ii. $A = B$ iff $\forall x \in E \mu_A(x) = \mu_B(x)$ and $v_A(x) = v_B(x)$

iii. $A^c = \left\{ \left(x, v_A(x), \mu_A(x) \right) : x \in E \right\}$.

iv. $A \cap B = \left\{ \left\{ x, \min\left(\mu_A(x), \mu_B(x) \right) \right\}, \max\left(v_A(x), v_B(x) \right) \right\} : x \in E$.

v. $A \cup B = \left\{ \left\{ x, \max\left(\mu_A(x), \mu_A(x) \right) \right\}, \min\left(v_A(x), v_B(x) \right) \right\} : x \in E$

Obviously every fuzzy set in an IFS has the form.

D.3 Research Methodology

In this present paper, researchers have showed a novel application of IFS in the number of cases and deaths due to various diseases for a particular year. Here, researchers have initially fuzzified the crisp data of the table showing the number of cases and deaths due to the diseases in the following years.

D.4 Euclidean Distance

D.4.1 The Normalized Euclidean Distance

$d_n - H(A,B)$ between two IFS.A and B is defined as follows:

$$d_n - H(A,B) = \left(\frac{1}{2n}\right)\left[\sum_{i=1}^{n}\left[\mu_A(x_i)-\mu_B(x_i)\right]^2 + \left[v_A(x_i)-v_B(x_i)\right]^2 \right.$$

$$\left. +\left[\pi_A(x_i)-\pi_B(x_i)\right]^2 \right]^{\frac{1}{2}}$$

where $X = \{x_1, x_2, \ldots, x_n\}$ for $i = 1, 2, \ldots, n$

Now if in an IFS table, each of the value sets represents a point in a 3D space which is of the form (x_i, y_i, z_i).

Then, the membership, non-membership, and the hesitation grades represent x_i, y_i and z_i, respectively. Now to calculate the distance between two categories, we are calculating the distances between the respective points associated with the categories.

This is done using the formula above, which is very similar to the general distance formula which we have used in analytical 3D geometry. Only what is different is that after calculating the distance, we are dividing it by 2n, i.e. the total number of points. The distance between any two points (x_1, y_1, z_1) and (x_2, y_2, z_2) in a 3D space is given by

$$\sqrt[2]{(x_1 - x_2)^2 + (y_1 - y_2)^2 + (z_1 - z_2)^2}$$

So, here we can use this concept to evaluate the distances between two value sets from two different IFS tables. And hence, get the distance between the two categories.

D.4.2 Sample Selection

In the given study, let D = {ACUTE DIARRHOEAL DISEASES, MALARIA, ACUTE RESPIRATORY INFECTION, JAPANESE ENCEPHALITIS, C} represent the set of diseases; Y = {2006, 2007, 2008, 2009, 2010} be the set of given years for which the number of cases and deaths occurred due to a particular disease; and S = {KERALA, KARNATAKA, MADHYA PRADESH, ANDHRA PRADESH, TAMIL NADU, WEST BENGAL} be the states that also show the number of cases and deaths occurred in that particular state due to a particular disease.

The intuitionistic fuzzy set is used as a tool with the membership degree μ (the degree of surety that a particular disease caused the given number of deaths), the non-membership degree v (the degree of surety that

TABLE D.2

Table Showing the Number of Deaths in Various Years

Diseases → Years ↓	Acute Diarrheal Diseases	Malaria	Acute Respiratory Infection	Japanese Encephalitis	Viral Hepatitis
2006	(0.48, 0.51, 0.01)	(0.003, 0.994, 0.003)	(0.86, 0.13, 0.01)	(0.1, 0.87, 0.01)	(0.1, 0.88, 0.02)
2007	(0.001, 0.99, 0.009)	(0.97, 0.02, 0.01)	(0.972, 0.02, 0.008)	(0.003, 0.995, 0.002)	(0.971, 0.021, 0.008)
2008	(0.74, 0.25, 0.01)	(0.968, 0.02, 0.012)	(0.965, 0.03, 0.005)	(0.98, 0.012, 0.008)	(0.976, 0.02, 0.004)
2009	(0.003, 0.99, 0.007)	(0.003, 0.986, 0.011)	(0.001, 0.986, 0.013)	(0.977, 0.02, 0.003)	(0.98, 0.009, 0.011)
2010	(0.01, 0.982, 0.017)	(0.003, 0.986, 0.011)	(0.001, 0.986, 0.013)	(0.977, 0.02, 0.003)	(0.98, 0.009, 0.011)

a particular disease did not cause the given number of deaths), and the hesitation degree π (the degree of unsurety of the number of deaths in a particular year).

Table D.2 shows the relationship between the numbers of deaths in the given years.

Table D.3 shows the membership, non-membership, and hesitation parts between the states and diseases:

Table D.4 is the final table that shows the membership, non-membership, and hesitation parts between States and Diseases.

TABLE D.3

Table Showing Membership, Non-Membership, and Hesitation Parts between States and Diseases

Diseases: States:	Acute Diarrheal Diseases	Malaria	Acute Respiratory Infection	Japanese Encephalitis	Viral Hepatitis
Kerala	(0.5, 0.43, 0.08)	(0.12, 0.86, 0.02)	(0.5, 0.46, 0.04)	(0.9, 0.08, 0.02)	(0.9, 0.18, 0.02)
Karnataka	(0.62, 0.35, 0.03)	(0.18, 0.81, 0.01)	(0.02, 0.95, 0.03)	(0.12, 0.87, 0.07)	(0.18, 0.81, 0.01)
Madhya Pradesh	(0.89, 0.10, 0.01)	(0.07, 0.90, 0.03)	(0.01, 0.98, 0.01)	(0.1, 0.87, 0.03)	(0.04, 0.95, 0.01)
Andhra Pradesh	(0.95, 0.04, 0.01)	(0.15, 0.84, 0.01)	(0.03, 0.95, 0.02)	(0.06, 0.93, 0.01)	(0.49, 0.5, 0.01)
Tamil Nadu	(0.48, 0.5, 0.02)	(0.17, 0.81, 0.02)	(0.02, 0.97, 0.01)	(0.15, 0.83, 0.02)	(0.88, 0.06, 0.06)
West Bengal	(0.97, 0.02, 0.01)	(0.96, 0.03, 0.01)	(0.98, 0.01, 0.01)	(0.98, 0.01, 0.01)	(0.95, 0.02, 0.03)

TABLE D.4

Table Showing Membership, Non-Membership, and Hesitation Parts between States and Diseases

Years: States:	2006	2007	2008	2009	2010
Kerala	(0.281)	(0.409)	(0.286)	(0.106)	(0.228)
Karnataka	(0.159)	(0.462)	(0.488)	(0.374)	(0.391)
Madhya Pradesh	(0.200)	(0.523)	(0.583)	(0.536)	(0.478)
Andhra Pradesh	(0.214)	(0.536)	(0.474)	(0.238)	(0.317)
Tamil Nadu	(0.266)	(0.482)	(0.521)	(0.493)	(0.530)
West Bengal	(0.467)	(0.254)	(0.327)	(0.458)	(0.469)

D.5 Conclusion

In this chapter, the normalized Euclidean distance method was used in IFS to determine the distribution of different states in India with the infection of different diseases like acute diarrheal diseases, malaria, acute respiratory infection, Japanese encephalitis, and varicella (c). At the end of the case study, we found that most of the infection occurs in West Bengal, Tamil Nadu, and Madhya Pradesh. A significant number of malaria patients are still present in these states.

There is a reasonable chance for the presence of a nonzero hesitation grade [h(x)] in Tables D.2 and D.3, which is predicting the confusion in the society or ancillary systems of approach.

Bibliography

Asplund, Kiell. "Clinometries in stroke research". *Stroke* 18, no. 2 (1987): 528–530.
Atanassov, Krassimir T. "Intuitionistic fuzzy sets". *Fuzzy Sets and Systems* 20 (1986): 87–96.
Atanassov, Krassimir, and Christo Georgeiv. "Intuitionistic fuzzy prolog". *Fuzzy Sets and Systems* 53 (1993): 121–128.
Djatna, Taufik. et al. "An intuitionistic fuzzy diagnosis analytics for stroke disease". *Journal of Big Data* 5 (2018): 35, 1–14.
Doubois, Didier, and Henri Prade. *Fuzzy Sets and Systems: Theory and Application.* Academic Press, New York (1980).
Kacprzyk, Janusz, and Eulalia. Szmidt. "Medical applications of intuitionistic fuzzy sets". *NIFS* 7, no. 4 (2001): 58–64.
Klaua, David J. "An analysis of programming techniques". *Teaching Machines and Programmed Learning II. Washington, DC: National Education Association* 7 (1965): 118–161.

Martino, R., N. Foley, and S. Bhogal. "Dysphasia after stroke, incidence and diagnosis". *Stroke* 36, no. 12 (2005): 2756–2763.

Meena, K., and K. V. Thomas. "An application of intuitionistic fuzzy sets in choice of discipline of study". *Global Journal of Pure and Applied Mathematics* 14, no. 6 (2018): 867–871.

Ottenbacher, Kenneth J., and Sylvie Jannell. "The results of clinical trials in stroke rehabilitation research". *Archives of Neurology* 50, no. 1 (1993): 37–44.

Robins, M., and H. M. Baum. "The national survey of stroke". *Stroke* 12, no. 2 (Suppl 1) (1981): 145–157.

Sanchez, Elie. "Solutions in composite fuzzy relation equations: application to medical diagnosis in Brouwerian logic". *Readings in Fuzzy Sets for Intelligent Systems, Morgan Kaufmann*, (1993) pp. 159–165..

Sullivan, Joe., and William H. Woodall. "A comparison of fuzzy forecasting and Markov modelling". *Fuzzy Sets and Systems* 64, no.3 (1994): 279–293.

Szmidt, Eulalia, and Janusz Kacprzyk. "Remarks on some applications of intuitionistic fuzzy sets in decision making". *Note on IFS* (1996).

Szmidt, Eulalia, Janusz Kacprzyk, and Paweł Bujnowski. "How to measure the amount of knowledge conveyed by Atanassov's intuitionistic fuzzy sets." Information Sciences (2014): 257, 276–285.

Thrift, Amanda G., et al. "Global stroke statistics". *International Journal of Stroke* 9, no. 1 (2014): 6–18.

Yu, Dejian, and Huchang Liao. "Visualization and quantitative research on intuitionistic fuzzy studies". *Journal of Intelligent and Fuzzy Systems* 30, no. 6 (2016): 3653–3663.

Zadeh, Lotfi A. "Fuzzy sets". *Information and Control* 8, no. 3 (1965): 338–353.

Zimmermann, Hans-Jurgen. *Fuzzy Set Theory and Applications* (2nd edition). Kluwer Academic Publishers, Amsterdam (1985).

Appendix E: Fuzzy Logic System Optimized

Green manufacturing is a basis of sustainable development strategy in machining industry that implies balancing between economical, ecological, and sociological segments of production. Green machining in manufacturing industry requires changes in the type and quantity of resources in the waste treatment, in the control of CO_2 emissions, and in the quantity of manufactured products. The main goal of machining process is to get high-quality products in a short time. In achieving that goal, cutting fluids and machining input parameters play a main role. The problems in the application of conventional cutting fluids during machining processes are human health and environmental pollution. The solution in terms of switching to green manufacturing is hidden in the application of alternative types of cooling, flushing, and lubricating techniques. These techniques are minimum quantity lubrication (MQL), cooling with cold compressed air (CCA), cryogenic cooling (CL) with different gasses, high-pressure cooling (HPC), minimum quantity lubrication and cooling (MQLC), near-dry machining (NDM), and dry machining by using new cutting tools and coatings. Dry machining is an environmentally friendly technique that is successfully applied in machining processes.

The most important advantages of dry machining are in reducing of disposal and cleaning costs of cutting fluids, reducing of environmental pollution, and no danger for health of operators. The elimination of CFs involves the loss of their positive effects, such as cooling, lubrication, and chip flushing. Development of production in terms of transition to dry machining is followed by development of new cutting tool materials. Advanced cutting tool materials and tool coatings are necessary during dry machining, but they are very expensive and increase the total machining costs. Some of these materials are sintered diamond, sintered CBN, ceramics (Al_2O_3), cermets, cemented carbides, etc. Coatings of cutting tools in dry machining replace the function of conventional cutting fluids in terms of decrease in friction and temperature in the cutting zone. The lubricating function of coatings in dry machining can be replaced with the soft coatings or so-called self-lubricating coatings like molybdenum disulfide (MoS_2) or tungsten carbide/carbon (WC/C).

In each machining process and even in dry machining, a surface roughness is one of the most used outputs that defines a quality of the final product. In this chapter, an investigation of the influence of input process parameters such as cutting speed, depth of cut, feed rate, and type of workpiece material on the surface roughness output was conducted. A combined approach of the Taguchi method and fuzzy logic technique was used to describe an influence of each process parameter on the surface roughness response and to define parameters values that lead to minimal surface roughness.

The Taguchi method is a simple and powerful tool for modeling, analysis, and optimization of the machining process. In this method, experimental data need to be transformed into signal-to-noise (S/N) ratio as the measure of the output quality characteristic. By the S/N ratio, it is possible to evaluate the effect of changing a particular input parameter on the analyzed process response. Depending on the criterion for the quality characteristic to be optimized, the S/N ratio can be divided into smaller-the-better, larger-the-better, and nominal-the-better. Regardless of the category of the process response, the larger S/N ratio corresponds to a better process performance characteristic. Accordingly, process parameter levels that lead to an optimal response have the highest S/N ratio values. Optimization of process response is performed by using the analysis of means (ANOM) and analysis of variance (ANOVA). The last step in the Taguchi optimization is conducting the confirmation experiment that should verify optimal settings of variable process parameters.

Fuzzy logic is an artificial intelligence method that is very useful for modeling complex processes where limited and imprecise information and numerical data do not allow development of accurate mathematical models by using classical methods such as regression analysis. In these cases, a fuzzy logic provides a way to better understand the process behavior by allowing the functional mapping between input and output observations. Each fuzzy system consists of four components: fuzzification module, fuzzy inference module, knowledge base, and defuzzification module. Fuzzification module converts numerical input data into linguistic variables by using different membership functions. There are various membership functions such as triangular, trapezoidal, and Gaussian. These functions define how each point in the input and output space is mapped to a degree of membership value between 0 and 1. Fuzzy inference module uses the knowledge base of membership functions and fuzzy IF–THEN rules to perform fuzzy reasoning and generate fuzzy linguistic output variables for corresponding inputs. Finally, the defuzzification module converts the aggregated fuzzy outputs into a non-fuzzy values (Figure E.1).

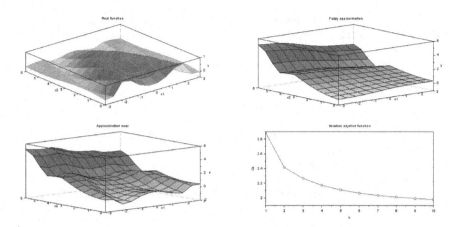

FIGURE E.1
Diagrammatic representation fuzzy logic structure.

```
SCILAB Output :
***** enters -qn code- (without bound cstr)
dimension=        12, epsq=  0.2220446049250313E-15,
verbosity level: imp=        3
max number of iterations allowed: iter=       100
max number of calls to costf allowed: nap=       100
--------------------------------------------------
iter num     1, nb calls=    1, f=  3.902
linear search: initial derivative=-0.7973
            step length= 0.1000E-01, df=-0.2231E-01,
derivative= 0.4327
iter num     2, nb calls=    2, f=  3.880
linear search: initial derivative=-0.3581
            step length= 0.1944E-01, df= 0.2753E-01,
derivative= 0.7763
            step length= 0.5878E-02, df=-0.6710E-02,
derivative=-0.1024E-03
iter num     3, nb calls=    4, f=  3.873
linear search: initial derivative=-0.3210
            step length= 0.1812E-02, df=-0.1292E-01,
derivative=-0.2968
            step length= 0.1812E-01, df=-0.7555E-01,
derivative=-0.1490E-01
iter num     4, nb calls=    6, f=  3.797
linear search: initial derivative=-0.2357
            step length= 0.8608E-01, df=-0.3810E-01,
derivative= 0.1229
iter num     5, nb calls=    7, f=  3.759
linear search: initial derivative=-0.1868
```

```
                step length= 0.7068E-01, df=-0.4072E-01,
derivative=-0.1228E-01
 iter num     6, nb calls=    8, f=   3.719
 linear search: initial derivative=-0.1770
                step length= 0.2907E-01, df=-0.7622E-01,
derivative=-0.1582
                step length= 0.2907    , df=   48.12     ,
derivative=  12.34
                step length= 0.5523E-01, df=-0.1458     ,
derivative=-0.1900
                step length= 0.1729    , df=   32.88     ,
derivative=  36.13
                step length= 0.6700E-01, df=-0.1840     ,
derivative=-0.2169
                step length= 0.1200    , df=   5.230     ,
derivative=  32.32
                step length= 0.9048E-01, df=-0.1876     ,
derivative= 0.7451
 iter num     7, nb calls=   15, f=   3.531
 linear search: initial derivative=-0.7510
                step length= 0.3847    , df= 0.4299E-01,
derivative= 0.2156
                step length= 0.1187    , df=-0.3265E-01,
derivative= 0.1070
 iter num     8, nb calls=   17, f=   3.498
 linear search: initial derivative=-0.3892
                step length= 0.2625E-01, df=-0.5315E-01,
derivative=-0.2457
 iter num     9, nb calls=   18, f=   3.445
 linear search: initial derivative=-0.2815
                step length= 0.2089    , df= 0.2686     ,
derivative=  1.660
                step length= 0.2861E-01, df=-0.7301E-02,
derivative=-0.1342E-02
 iter num    10, nb calls=   20, f=   3.438
 linear search: initial derivative=-0.2979
                step length= 0.2977E-02, df=-0.1411E-01,
derivative=-0.2778
                step length= 0.2977E-01, df=-0.8258E-01,
derivative= 0.6900E-02
 iter num    11, nb calls=   22, f=   3.355
 linear search: initial derivative=-0.6375
                step length= 0.1890    , df=-0.1194     ,
derivative=-0.2513
 iter num    12, nb calls=   23, f=   3.236
 linear search: initial derivative=-0.4161
                step length= 0.7462E-01, df=   4.558     ,
derivative=  20.00
```

```
                step length= 0.7462E-02, df=-0.1006E-01,
derivative= 0.8691E-01
 iter num    13, nb calls=  25, f=  3.226
 linear search: initial derivative=-0.3428
                step length= 0.4977E-01, df=-0.1838E-01,
derivative=-0.2827
                step length= 0.2775   , df=-0.4733E-01,
derivative= 0.1124
 iter num    14, nb calls=  27, f=  3.179
 linear search: initial derivative=-0.3438
                step length= 0.2447E-01, df=-0.7054E-01,
derivative=-0.1530
 iter num    15, nb calls=  28, f=  3.108
 linear search: initial derivative=-0.3940
                step length= 0.7066   , df=-0.8333E-01,
derivative=-0.7593E-01
 iter num    16, nb calls=  29, f=  3.025
 linear search: initial derivative=-0.2941
                step length= 0.6901   , df=  1.143   ,
derivative=  10.01
                step length= 0.3004   , df=-0.1205E-01,
derivative= 0.4206
 iter num    17, nb calls=  31, f=  3.013
 linear search: initial derivative=-0.6662
                step length= 0.1245   , df=-0.2052E-01,
derivative=-0.4694
                step length= 0.4252   , df=-0.4174E-01,
derivative=-0.2227E-01
 iter num    18, nb calls=  33, f=  2.971
 linear search: initial derivative=-0.3877
                step length= 0.1929   , df=-0.3468E-01,
derivative= 0.6024E-01
 iter num    19, nb calls=  34, f=  2.936
 linear search: initial derivative=-0.1041
                step length= 1.0000   , df=-0.3393E-01,
derivative=-0.3296E-01
 iter num    20, nb calls=  35, f=  2.902
 linear search: initial derivative=-0.8133E-01
                step length= 1.0000   , df= 0.6779E-04,
derivative= 0.1273
                step length= 0.5517   , df=-0.7890E-02,
derivative=-0.5455E-03
 iter num    21, nb calls=  37, f=  2.894
 linear search: initial derivative=-0.5008E-01
                step length= 1.0000   , df=-0.6422E-03,
derivative= 0.3584E-01
                step length= 0.5161   , df=-0.2389E-02,
derivative=-0.8397E-04
```

```
iter num    22, nb calls=  39, f=  2.892
linear search: initial derivative=-0.4973E-01
                step length= 0.9318    , df=-0.3089E-02,
derivative=-0.1430E-01
iter num    23, nb calls=  40, f=  2.889
linear search: initial derivative=-0.3527E-01
                step length= 0.9293    , df= 0.7432E-03,
derivative= 0.5645E-01
                step length= 0.4537    , df=-0.1588E-02,
derivative= 0.1075E-03
iter num    24, nb calls=  42, f=  2.887
linear search: initial derivative=-0.3520E-01
                step length= 0.7799    , df=-0.1175E-02,
derivative= 0.8757E-02
iter num    25, nb calls=  43, f=  2.886
linear search: initial derivative=-0.2633E-01
                step length= 1.0000    , df=-0.2796E-03,
derivative= 0.1621E-01
iter num    26, nb calls=  44, f=  2.886
linear search: initial derivative=-0.1799E-01
                step length= 0.9246    , df=-0.1580E-03,
derivative= 0.7526E-02
iter num    27, nb calls=  45, f=  2.886
linear search: initial derivative=-0.1396E-01
                step length= 0.3774    , df=-0.4243E-04,
derivative= 0.9698E-02
iter num    28, nb calls=  46, f=  2.886
linear search: initial derivative=-0.7320E-02
                step length= 0.4870    , df=-0.6510E-04,
derivative=-0.3884E-02
iter num    29, nb calls=  47, f=  2.886
linear search: initial derivative=-0.3596E-02
                step length= 1.0000    , df=-0.3113E-04,
derivative= 0.1182E-03
iter num    30, nb calls=  48, f=  2.886
linear search: initial derivative=-0.4634E-02
                step length= 1.0000    , df=-0.3293E-05,
derivative=-0.1753E-02
iter num    31, nb calls=  49, f=  2.886
linear search: initial derivative=-0.2429E-02
                step length= 1.0000    , df=-0.3592E-05,
derivative=-0.8909E-03
iter num    32, nb calls=  50, f=  2.886
linear search: initial derivative=-0.9950E-03
                step length= 1.0000    , df=-0.1174E-05,
derivative=-0.1643E-03
iter num    33, nb calls=  51, f=  2.886
linear search: initial derivative=-0.6381E-03
                step length= 1.0000    , df=-0.1187E-06,
derivative=-0.1349E-03
```

```
iter num    34, nb calls=  52, f=  2.886
linear search: initial derivative=-0.2095E-03
               step length= 1.0000    , df=-0.1212E-07,
derivative=-0.9586E-05
iter num    35, nb calls=  53, f=  2.886
linear search: initial derivative=-0.1236E-04
               step length= 1.0000    , df=-0.2897E-09,
derivative= 0.1611E-05
iter num    36, nb calls=  54, f=  2.886
linear search: initial derivative=-0.1002E-04
               step length= 1.0000    , df=-0.1664E-10,
derivative= 0.4531E-06
iter num    37, nb calls=  55, f=  2.886
linear search: initial derivative=-0.1049E-05
               step length= 1.0000    , df=-0.1670E-12,
derivative= 0.1301E-06
iter num    38, nb calls=  56, f=  2.886
linear search: initial derivative=-0.1178E-06
               step length= 1.0000    , df=-0.7105E-14,
derivative=-0.1196E-08
iter num    39, nb calls=  57, f=  2.886
linear search: initial derivative=-0.1087E-07
               step length= 1.0000    , df= 0.0000E+00,
derivative=-0.9976E-09
               step length= 0.3184    , df= 0.0000E+00,
derivative=-0.7640E-08
               step length= 0.7607E-01, df= 0.0000E+00,
derivative=-0.1008E-07
               step length= 0.1654E-01, df=-0.4441E-15,
derivative=-0.1070E-07
               step length= 0.3309E-01, df= 0.4441E-15,
derivative=-0.1047E-07
               step length= 0.1820E-01, df= 0.0000E+00,
derivative=-0.1068E-07
               step length= 0.1671E-01, df=-0.4441E-15,
derivative=-0.1066E-07
               step length= 0.1745E-01, df= 0.0000E+00,
derivative=-0.1060E-07
               step length= 0.1678E-01, df= 0.0000E+00,
derivative=-0.1060E-07
               step length= 0.1672E-01, df= 0.4441E-15,
derivative=-0.1076E-07
               step length= 0.1671E-01, df= 0.8882E-15,
derivative=-0.1072E-07
               step length= 0.1671E-01, df= 0.0000E+00,
derivative=-0.1065E-07
               step length= 0.1671E-01, df= 0.0000E+00,
derivative=-0.1066E-07
               step length= 0.1671E-01, df=-0.4441E-15,
derivative=-0.1066E-07
```

```
                    step length= 0.1671E-01, df=-0.4441E-15,
derivative=-0.1066E-07
                    step length= 0.1671E-01, df= 0.0000E+00,
derivative=-0.1067E-07
                    step length= 0.1671E-01, df=-0.4441E-15,
derivative=-0.1066E-07
                    step length= 0.1671E-01, df= 0.0000E+00,
derivative=-0.1067E-07
 iter num    40, nb calls=  75, f=  2.886
 linear search: initial derivative=-0.1074E-07
                    step length= 1.0000    , df= 0.4441E-15,
derivative=-0.1026E-08
                    step length= 0.1000E+00, df= 0.4441E-15,
derivative=-0.9639E-08
                    step length= 0.1000E-01, df= 0.8882E-15,
derivative=-0.1055E-07
                    step length= 0.1000E-02, df= 0.4441E-15,
derivative=-0.1084E-07
                    step length= 0.1000E-03, df= 0.4441E-15,
derivative=-0.1064E-07
                    step length= 0.1000E-04, df= 0.4441E-15,
derivative=-0.1073E-07
                    step length= 0.1000E-05, df= 0.4441E-15,
derivative=-0.1075E-07
                    step length= 0.1000E-06, df= 0.0000E+00,
derivative=-0.1075E-07
                    step length= 0.2113E-07, df= 0.4441E-15,
derivative=-0.1079E-07
                    step length= 0.2113E-08, df= 0.0000E+00,
derivative=-0.1074E-07
                    step length= 0.4465E-09, df= 0.0000E+00,
derivative=-0.1074E-07
 iter num    41, nb calls=  86, f=  2.886
 linear search: initial derivative=-0.1087E-07
                    step length= 1.0000    , df= 0.8882E-15,
derivative=-0.9839E-09
                    step length= 0.1000E+00, df= 0.4441E-15,
derivative=-0.9820E-08
                    step length= 0.1000E-01, df= 0.1332E-14,
derivative=-0.1072E-07
                    step length= 0.1000E-02, df= 0.4441E-15,
derivative=-0.1083E-07
                    step length= 0.1000E-03, df= 0.4441E-15,
derivative=-0.1075E-07
                    step length= 0.1000E-04, df= 0.4441E-15,
derivative=-0.1087E-07
                    step length= 0.1000E-05, df= 0.4441E-15,
derivative=-0.1094E-07
                    step length= 0.1000E-06, df= 0.4441E-15,
derivative=-0.1088E-07
```

```
              step length= 0.1000E-07, df=-0.4441E-15,
derivative=-0.1085E-07
              step length= 0.2000E-07, df= 0.4441E-15,
derivative=-0.1094E-07
              step length= 0.1100E-07, df=-0.4441E-15,
derivative=-0.1085E-07
              step length= 0.1550E-07, df= 0.4441E-15,
derivative=-0.1090E-07
              step length= 0.1145E-07, df=-0.4441E-15,
derivative=-0.1085E-07
              step length= 0.1348E-07, df= 0.4441E-15,
derivative=-0.1087E-07
   iter num    41, nb calls= 100, f=  2.886
***** leaves -qn code-, gradient norm=  0.9493375269080503E-07
Optim stops: maximum number of calls to f is reached.

   Iteration #1 calculation time =8.244 objetive function =
2.8855312

***** enters -qn code- (without bound cstr)
dimension=        12, epsq=  0.2220446049250313E-15, verbosity
level: imp=       3
max number of iterations allowed: iter=        100
max number of calls to costf allowed: nap=       100
-------------------------------------------------
   iter num     1, nb calls=  1, f=  2.526
   linear search: initial derivative= -7.795
              step length= 0.1000E-01, df=  7.307    ,
derivative=  69.67
              step length= 0.1000E-02, df=-0.8927E-01,
derivative= -2.043
   iter num     2, nb calls=  3, f=  2.436
   linear search: initial derivative= -2.225
              step length= 0.1497E-01, df=  6.611    ,
derivative=  171.3
              step length= 0.1497E-02, df= 0.4672E-01,
derivative=  13.92
              step length= 0.2080E-03, df=-0.1241E-02,
derivative=-0.7121E-04
   iter num     3, nb calls=  6, f=  2.435
   linear search: initial derivative=-0.8072
              step length= 0.1507E-02, df=-0.1932E-02,
derivative=-0.4498
   iter num     4, nb calls=  7, f=  2.433
   linear search: initial derivative=-0.5808
              step length= 0.2551E-02, df=-0.3410E-02,
derivative=-0.4438
              step length= 0.1078E-01, df=-0.8081E-02,
derivative= 0.1239E-01
   iter num     5, nb calls=  9, f=  2.425
```

```
linear search: initial derivative=-0.1614
                step length= 0.2956    , df= 0.7320E-02,
derivative= 0.3781
                step length= 0.1161    , df=-0.3250E-02,
derivative= 0.2794E-02
 iter num    6, nb calls=  11, f=  2.422
 linear search: initial derivative=-0.4803E-01
                step length= 1.0000    , df= 0.4134E-02,
derivative= 0.1565
                step length= 0.2333    , df=-0.4218E-03,
derivative= 0.2234E-03
 iter num    7, nb calls=  13, f=  2.422
 linear search: initial derivative=-0.4300E-01
                step length= 0.8814E-01, df=-0.2105E-03,
derivative= 0.2139E-01
 iter num    8, nb calls=  14, f=  2.421
 linear search: initial derivative=-0.4293E-01
                step length= 0.8976E-03, df=-0.3854E-03,
derivative=-0.3570E-01
                step length= 0.5352E-02, df=-0.1276E-02,
derivative=-0.1242E-02
 iter num    9, nb calls=  16, f=  2.420
 linear search: initial derivative=-0.2074E-01
                step length= 1.0000    , df=-0.5369E-03,
derivative=-0.8935E-02
 iter num   10, nb calls=  17, f=  2.420
 linear search: initial derivative=-0.1242E-01
                step length= 1.0000    , df= 0.3858E-02,
derivative= 0.1042      ·
                step length= 0.1095    , df=-0.5767E-04,
derivative= 0.6665E-04
 iter num   11, nb calls=  19, f=  2.419
 linear search: initial derivative=-0.1052E-01
                step length= 0.7455E-01, df=-0.1074E-03,
derivative=-0.9066E-02
                step length= 0.5334    , df=-0.3754E-03,
derivative= 0.1772E-02
 iter num   12, nb calls=  21, f=  2.419
 linear search: initial derivative=-0.1125E-01
                step length= 0.5168    , df= 0.1358E-03,
derivative= 0.1532E-01
                step length= 0.2188    , df=-0.1586E-03,
derivative= 0.6251E-05
 iter num   13, nb calls=  23, f=  2.419
 linear search: initial derivative=-0.1432E-01
                step length= 0.3561    , df=-0.2905E-03,
derivative=-0.1184E-01
                step length= 1.989     , df=-0.5057E-03,
derivative= 0.1326E-01
 iter num   14, nb calls=  25, f=  2.418
```

```
linear search: initial derivative=-0.4138E-01
                step length= 0.7565    , df=-0.5478E-03,
derivative=-0.4889E-02
 iter num    15, nb calls=  26, f=   2.418
 linear search: initial derivative=-0.5335E-02
                step length= 1.0000    , df=-0.1843E-03,
derivative= 0.1057E-02
 iter num    16, nb calls=  27, f=   2.418
 linear search: initial derivative=-0.9784E-02
                step length= 1.0000    , df=-0.1374E-03,
derivative=-0.2718E-02
 iter num    17, nb calls=  28, f=   2.418
 linear search: initial derivative=-0.2879E-02
                step length= 1.0000    , df=-0.5889E-04,
derivative=-0.6648E-03
 iter num    18, nb calls=  29, f=   2.418
 linear search: initial derivative=-0.1994E-02
                step length= 1.0000    , df=-0.1266E-04,
derivative=-0.1658E-03
 iter num    19, nb calls=  30, f=   2.418
 linear search: initial derivative=-0.4405E-03
                step length= 1.0000    , df=-0.7589E-06,
derivative= 0.3310E-03
 iter num    20, nb calls=  31, f=   2.418
 linear search: initial derivative=-0.4690E-03
                step length= 0.7891    , df=-0.8877E-06,
derivative=-0.8313E-04
 iter num    21, nb calls=  32, f=   2.418
 linear search: initial derivative=-0.1116E-03
                step length= 1.0000    , df=-0.3386E-07,
derivative=-0.5082E-05
 iter num    22, nb calls=  33, f=   2.418
 linear search: initial derivative=-0.1390E-04
                step length= 1.0000    , df=-0.3300E-09,
derivative=-0.3244E-05
 iter num    23, nb calls=  34, f=   2.418
 linear search: initial derivative=-0.6600E-05
                step length= 1.0000    , df=-0.3175E-10,
derivative=-0.2882E-06
 iter num    24, nb calls=  35, f=   2.418
 linear search: initial derivative=-0.1139E-05
                step length= 1.0000    , df=-0.2753E-12,
derivative=-0.5776E-07
 iter num    25, nb calls=  36, f=   2.418
 linear search: initial derivative=-0.8738E-07
                step length= 1.0000    , df=-0.6661E-14,
derivative= 0.1404E-08
 iter num    26, nb calls=  37, f=   2.418
 linear search: initial derivative=-0.5640E-08
```

```
              step length= 1.0000    , df= 0.4441E-15,
derivative=-0.2154E-09
              step length= 0.1000E+00, df= 0.8882E-15,
derivative=-0.5093E-08
              step length= 0.1000E-01, df= 0.0000E+00,
derivative=-0.5619E-08
              step length= 0.2116E-02, df=-0.4441E-15,
derivative=-0.5618E-08
              step length= 0.4233E-02, df= 0.8882E-15,
derivative=-0.5657E-08
              step length= 0.2328E-02, df= 0.0000E+00,
derivative=-0.5553E-08
              step length= 0.2137E-02, df=-0.4441E-15,
derivative=-0.5558E-08
              step length= 0.2233E-02, df= 0.4441E-15,
derivative=-0.5467E-08
              step length= 0.2147E-02, df=-0.1332E-14,
derivative=-0.5439E-08
              step length= 0.2190E-02, df=-0.1332E-14,
derivative=-0.5679E-08
              step length= 0.2211E-02, df=-0.4441E-15,
derivative=-0.5554E-08
              step length= 0.2222E-02, df=-0.4441E-15,
derivative=-0.5718E-08
              step length= 0.2227E-02, df= 0.0000E+00,
derivative=-0.5707E-08
              step length= 0.2223E-02, df=-0.4441E-15,
derivative=-0.5574E-08
              step length= 0.2225E-02, df= 0.0000E+00,
derivative=-0.5435E-08
              step length= 0.2223E-02, df=-0.4441E-15,
derivative=-0.5522E-08
              step length= 0.2224E-02, df=-0.4441E-15,
derivative=-0.5635E-08
              step length= 0.2224E-02, df=-0.1332E-14,
derivative=-0.5508E-08
              step length= 0.2225E-02, df= 0.0000E+00,
derivative=-0.5498E-08
              step length= 0.2224E-02, df=-0.1332E-14,
derivative=-0.5556E-08
              step length= 0.2225E-02, df=-0.4441E-15,
derivative=-0.5492E-08
              step length= 0.2225E-02, df= 0.0000E+00,
derivative=-0.5405E-08
              step length= 0.2225E-02, df=-0.4441E-15,
derivative=-0.5490E-08
              step length= 0.2225E-02, df=-0.1332E-14,
derivative=-0.5589E-08
```

```
                step length= 0.2225E-02, df=-0.4441E-15,
derivative=-0.5443E-08
                step length= 0.2225E-02, df=-0.4441E-15,
derivative=-0.5447E-08
                step length= 0.2225E-02, df= 0.0000E+00,
derivative=-0.5441E-08
                step length= 0.2225E-02, df=-0.4441E-15,
derivative=-0.5447E-08
                step length= 0.2225E-02, df=-0.4441E-15,
derivative=-0.5418E-08
                step length= 0.2225E-02, df=-0.4441E-15,
derivative=-0.5418E-08
                step length= 0.2225E-02, df= 0.0000E+00,
derivative=-0.5441E-08
 iter num    27, nb calls=  68, f=  2.418
 linear search: initial derivative=-0.5655E-08
                step length= 1.0000    , df= 0.2665E-14,
derivative=-0.8798E-10
                step length= 0.1000E+00, df=-0.4441E-15,
derivative=-0.5030E-08
                step length= 0.2000    , df= 0.4441E-15,
derivative=-0.4563E-08
                step length= 0.1100    , df= 0.2220E-14,
derivative=-0.5156E-08
                step length= 0.1010    , df= 0.4441E-15,
derivative=-0.5134E-08
                step length= 0.1001    , df= 0.8882E-15,
derivative=-0.4989E-08
                step length= 0.1000    , df= 0.8882E-15,
derivative=-0.5125E-08
                step length= 0.1000E+00, df= 0.8882E-15,
derivative=-0.5046E-08
                step length= 0.1000E+00, df= 0.0000E+00,
derivative=-0.5059E-08
                step length= 0.1000E+00, df= 0.0000E+00,
derivative=-0.5062E-08
                step length= 0.1000E+00, df=-0.4441E-15,
derivative=-0.5030E-08
                step length= 0.1000E+00, df= 0.0000E+00,
derivative=-0.5062E-08
                step length= 0.1000E+00, df=-0.4441E-15,
derivative=-0.5030E-08
                step length= 0.1000E+00, df= 0.0000E+00,
derivative=-0.5018E-08
                step length= 0.1000E+00, df=-0.4441E-15,
derivative=-0.5030E-08
                step length= 0.1000E+00, df=-0.4441E-15,
derivative=-0.5030E-08
```

```
                step length= 0.1000E+00, df=-0.4441E-15,
derivative=-0.5030E-08
                step length= 0.1000E+00, df=-0.4441E-15,
derivative=-0.5030E-08
                step length= 0.1000E+00, df= 0.0000E+00,
derivative=-0.5018E-08
 iter num    28, nb calls=  87, f=  2.418
 linear search: initial derivative=-0.5241E-08
                step length= 1.0000     , df= 0.3109E-14,
derivative=-0.4631E-09
                step length= 0.1000E+00, df= 0.1332E-14,
derivative=-0.4872E-08
                step length= 0.1000E-01, df= 0.3109E-14,
derivative=-0.5341E-08
                step length= 0.1000E-02, df= 0.2220E-14,
derivative=-0.5308E-08
                step length= 0.1000E-03, df= 0.1776E-14,
derivative=-0.5329E-08
                step length= 0.1000E-04, df= 0.1332E-14,
derivative=-0.5349E-08
                step length= 0.1000E-05, df= 0.2220E-14,
derivative=-0.5368E-08
                step length= 0.1000E-06, df= 0.4441E-15,
derivative=-0.5286E-08
                step length= 0.1000E-07, df=-0.4441E-15,
derivative=-0.5187E-08
                step length= 0.2000E-07, df= 0.8882E-15,
derivative=-0.5220E-08
                step length= 0.1100E-07, df= 0.1332E-14,
derivative=-0.5277E-08
                step length= 0.1010E-07, df=-0.4441E-15,
derivative=-0.5187E-08
                step length= 0.1055E-07, df= 0.1332E-14,
derivative=-0.5328E-08
 iter num    29, nb calls= 100, f=  2.418
 linear search: initial derivative=-0.5378E-08
 iter num    29, nb calls= 100, f=  2.418
***** leaves -qn code-, gradient norm=  0.2256490735931721E-06
Optim stops: maximum number of calls to f is reached.

  Iteration #2 calculation time =5.14 objetive function =
2.4175037

***** enters -qn code- (without bound cstr)
dimension=        12, epsq=  0.2220446049250313E-15, verbosity
level: imp=         3
max number of iterations allowed: iter=          100
max number of calls to costf allowed: nap=         100
--------------------------------------------------
```

```
iter num     1, nb calls=   1, f=   2.330
linear search: initial derivative= -8.983
              step length= 0.1000E-01, df=  9.588    ,
derivative=  51.94
              step length= 0.1000E-02, df=-0.4482E-01,
derivative=  5.073
iter num     2, nb calls=   3, f=   2.285
linear search: initial derivative= -1.658
              step length= 0.1561E-01, df=  5.208    ,
derivative=  198.9
              step length= 0.1561E-02, df= 0.4157E-01,
derivative=  17.09
              step length= 0.1561E-03, df=-0.3937E-03,
derivative= 0.2021
iter num     3, nb calls=   6, f=   2.285
linear search: initial derivative=-0.4637
              step length= 0.1830E-02, df=-0.5847E-03,
derivative=-0.2249
iter num     4, nb calls=   7, f=   2.284
linear search: initial derivative=-0.3751
              step length= 0.1817E-02, df=-0.1081E-02,
derivative=-0.3185
              step length= 0.1204E-01, df=-0.3888E-02,
derivative=-0.2367E-02
iter num     5, nb calls=   9, f=   2.280
linear search: initial derivative=-0.2599
              step length= 0.6320E-01, df=-0.4441E-02,
derivative=-0.3423E-01
iter num     6, nb calls=  10, f=   2.276
linear search: initial derivative=-0.2347
              step length= 0.1627    , df=-0.3671E-02,
derivative= 0.4410E-01
iter num     7, nb calls=  11, f=   2.272
linear search: initial derivative=-0.2227
              step length= 0.2281    , df= 0.1407    ,
derivative=  8.903
              step length= 0.2281E-01, df= 0.7049E-03,
derivative= 0.6518
              step length= 0.5830E-02, df=-0.9386E-04,
derivative=-0.4203E-08
iter num     8, nb calls=  14, f=   2.272
linear search: initial derivative=-0.2344
              step length= 0.9888E-03, df=-0.1865E-03,
derivative=-0.2313
              step length= 0.9888E-02, df=-0.1752E-02,
derivative=-0.2028
              step length= 0.7256E-01, df=-0.6354E-02,
derivative= 0.3239E-01
iter num     9, nb calls=  17, f=   2.266
```

```
linear search: initial derivative=-0.2302E-01
               step length= 1.0000    , df=-0.1906E-03,
derivative= 0.7142E-02
 iter num   10, nb calls=  18, f=  2.265
 linear search: initial derivative=-0.1254E-01
               step length= 1.0000    , df=-0.8362E-04,
derivative= 0.9505E-03
 iter num   11, nb calls=  19, f=  2.265
 linear search: initial derivative=-0.6604E-02
               step length= 1.0000    , df= 0.6456E-03,
derivative= 0.9370E-01
               step length= 0.1000E+00, df=-0.2394E-05,
derivative= 0.3383E-02
 iter num   12, nb calls=  21, f=  2.265
 linear search: initial derivative=-0.5456E-02
               step length= 0.9262E-01, df=-0.4524E-05,
derivative=-0.4854E-02
               step length= 0.8396    , df=-0.2169E-04,
derivative= 0.3708E-05
 iter num   13, nb calls=  23, f=  2.265
 linear search: initial derivative=-0.4126E-02
               step length= 0.4100    , df= 0.2091E-05,
derivative= 0.4496E-02
               step length= 0.1952    , df=-0.1031E-04,
derivative= 0.7545E-07
 iter num   14, nb calls=  25, f=  2.265
 linear search: initial derivative=-0.2485E-02
               step length= 0.7406    , df=-0.1784E-04,
derivative=-0.1812E-02
               step length=  2.720    , df=-0.3708E-04,
derivative= 0.1223E-03
 iter num   15, nb calls=  27, f=  2.265
 linear search: initial derivative=-0.3025E-02
               step length= 1.0000    , df=-0.5258E-05,
derivative=-0.2550E-03
 iter num   16, nb calls=  28, f=  2.265
 linear search: initial derivative=-0.3817E-03
               step length= 1.0000    , df=-0.3510E-06,
derivative=-0.2146E-04
 iter num   17, nb calls=  29, f=  2.265
 linear search: initial derivative=-0.8015E-04
               step length= 1.0000    , df=-0.9134E-08,
derivative=-0.6934E-05
 iter num   18, nb calls=  30, f=  2.265
 linear search: initial derivative=-0.1235E-04
               step length= 1.0000    , df=-0.1534E-09,
derivative=-0.4150E-06
 iter num   19, nb calls=  31, f=  2.265
 linear search: initial derivative=-0.1524E-05
```

```
              step length= 1.0000    , df=-0.1296E-11,
derivative= 0.5930E-07
 iter num    20, nb calls=  32, f=   2.265
 linear search: initial derivative=-0.9542E-07
              step length= 1.0000    , df=-0.6217E-14,
derivative= 0.1792E-08
              step length= 0.3840    , df=-0.1465E-13,
derivative=-0.5802E-07
 iter num    21, nb calls=  34, f=   2.265
 linear search: initial derivative=-0.6228E-07
              step length= 0.7644    , df= 0.4441E-15,
derivative=-0.1506E-07
              step length= 0.2198    , df=-0.1776E-14,
derivative=-0.4870E-07
              step length= 0.4921    , df=-0.1332E-14,
derivative=-0.3195E-07
              step length= 0.2811    , df=-0.2220E-14,
derivative=-0.4499E-07
              step length= 0.3866    , df=-0.2220E-14,
derivative=-0.3844E-07
 iter num    22, nb calls=  39, f=   2.265
 linear search: initial derivative=-0.3841E-07
              step length= 0.3006    , df= 0.8882E-15,
derivative=-0.2665E-07
              step length= 0.5535E-01, df= 0.4441E-15,
derivative=-0.3616E-07
              step length= 0.6912E-02, df= 0.0000E+00,
derivative=-0.3809E-07
              step length= 0.1465E-02, df=-0.4441E-15,
derivative=-0.3831E-07
              step length= 0.2931E-02, df=-0.8882E-15,
derivative=-0.3840E-07
              step length= 0.4921E-02, df= 0.8882E-15,
derivative=-0.3825E-07
              step length= 0.3130E-02, df= 0.4441E-15,
derivative=-0.3829E-07
              step length= 0.2951E-02, df= 0.8882E-15,
derivative=-0.3818E-07
              step length= 0.2933E-02, df= 0.0000E+00,
derivative=-0.3818E-07
              step length= 0.2931E-02, df=-0.8882E-15,
derivative=-0.3825E-07
              step length= 0.2932E-02, df= 0.0000E+00,
derivative=-0.3829E-07
              step length= 0.2931E-02, df=-0.8882E-15,
derivative=-0.3831E-07
              step length= 0.2932E-02, df= 0.8882E-15,
derivative=-0.3817E-07
```

```
              step length= 0.2931E-02, df= 0.0000E+00,
derivative=-0.3822E-07
              step length= 0.2931E-02, df= 0.0000E+00,
derivative=-0.3827E-07
              step length= 0.2931E-02, df=-0.4441E-15,
derivative=-0.3830E-07
              step length= 0.2931E-02, df=-0.1776E-14,
derivative=-0.3827E-07
              step length= 0.2931E-02, df=-0.8882E-15,
derivative=-0.3821E-07
              step length= 0.2931E-02, df=-0.4441E-15,
derivative=-0.3826E-07
              step length= 0.2931E-02, df= 0.0000E+00,
derivative=-0.3822E-07
              step length= 0.2931E-02, df=-0.4441E-15,
derivative=-0.3826E-07
              step length= 0.2931E-02, df=-0.4441E-15,
derivative=-0.3826E-07
              step length= 0.2931E-02, df= 0.0000E+00,
derivative=-0.3822E-07
 iter num   23, nb calls= 62, f=  2.265
 linear search: initial derivative=-0.3831E-07
              step length= 0.3027    , df= 0.2665E-14,
derivative=-0.2671E-07
              step length= 0.3835E-01, df= 0.2220E-14,
derivative=-0.3679E-07
              step length= 0.3835E-02, df= 0.1776E-14,
derivative=-0.3811E-07
              step length= 0.3835E-03, df= 0.1776E-14,
derivative=-0.3823E-07
              step length= 0.3835E-04, df= 0.2220E-14,
derivative=-0.3823E-07
              step length= 0.3835E-05, df= 0.2220E-14,
derivative=-0.3814E-07
              step length= 0.3835E-06, df= 0.8882E-15,
derivative=-0.3815E-07
              step length= 0.3835E-07, df= 0.3553E-14,
derivative=-0.3825E-07
              step length= 0.3835E-08, df= 0.1776E-14,
derivative=-0.3836E-07
              step length= 0.3835E-09, df= 0.0000E+00,
derivative=-0.3831E-07
              step length= 0.8104E-10, df= 0.0000E+00,
derivative=-0.3831E-07
 iter num   23, nb calls=  73, f=  2.265
***** leaves -qn code-, gradient norm=  0.1475063153491663E-05
End of optimization.

   Iteration #3 calculation time =3.581 objetive function =
2.2652874
```

```
***** enters -qn code- (without bound cstr)
dimension=        12, epsq=  0.2220446049250313E-15, verbosity
level: imp=        3
max number of iterations allowed: iter=          100
max number of calls to costf allowed: nap=        100
---------------------------------------------------
 iter num     1, nb calls=    1, f=  2.213
 linear search: initial derivative= -7.750
                step length= 0.1000E-01, df=  9.947      ,
derivative=  50.31
                step length= 0.1000E-02, df= 0.1512E-01,
derivative=  10.32
                step length= 0.4679E-03, df=-0.3472E-01,
derivative=-0.4150E-02
 iter num     2, nb calls=    4, f=  2.178
 linear search: initial derivative= -1.361
                step length= 0.1514E-01, df=  4.603      ,
derivative=  185.8
                step length= 0.1514E-02, df= 0.3774E-01,
derivative=  16.20
                step length= 0.1514E-03, df=-0.2498E-03,
derivative= 0.3824
 iter num     3, nb calls=    7, f=  2.178
 linear search: initial derivative=-0.4277
                step length= 0.1026E-02, df=-0.4020E-03,
derivative=-0.2607
 iter num     4, nb calls=    8, f=  2.177
 linear search: initial derivative=-0.3318
                step length= 0.1211E-02, df=-0.7531E-03,
derivative=-0.2898
                step length= 0.9561E-02, df=-0.3176E-02,
derivative=-0.1626E-03
 iter num     5, nb calls=   10, f=  2.174
 linear search: initial derivative=-0.1494
                step length= 0.9533E-01, df= 0.2853E-02,
derivative= 0.3035
                step length= 0.3434E-01, df=-0.1162E-02,
derivative= 0.6881E-04
 iter num     6, nb calls=   12, f=  2.173
 linear search: initial derivative=-0.6501E-01
                step length= 0.1973     , df=-0.6615E-03,
derivative= 0.2748E-01
 iter num     7, nb calls=   13, f=  2.172
 linear search: initial derivative=-0.4297E-01
                step length= 0.2942     , df= 0.1817      ,
derivative=  11.71
                step length= 0.2942E-01, df= 0.1731E-02,
derivative=  1.166
                step length= 0.2942E-02, df= 0.5430E-05,
derivative= 0.7824E-01
```

```
               step length= 0.1043E-02, df=-0.2345E-05,
derivative=-0.7623E-07
 iter num    8, nb calls=  17, f=  2.172
 linear search: initial derivative=-0.3710E-01
               step length= 0.1574E-02, df=-0.4678E-05,
derivative=-0.3692E-01
               step length= 0.1574E-01, df=-0.4573E-04,
derivative=-0.3525E-01
               step length= 0.1574    , df=-0.3505E-03,
derivative=-0.1820E-01
 iter num    9, nb calls=  20, f=  2.172
 linear search: initial derivative=-0.2977E-01
               step length= 0.6472    , df=-0.2761E-03,
derivative= 0.5847E-02
 iter num   10, nb calls=  21, f=  2.172
 linear search: initial derivative=-0.1010E-01
               step length= 1.0000    , df=-0.2975E-04,
derivative= 0.4285E-02
 iter num   11, nb calls=  22, f=  2.172
 linear search: initial derivative=-0.9478E-02
               step length= 0.8773E-01, df=-0.9940E-06,
derivative= 0.9257E-02
               step length= 0.4472E-01, df=-0.1520E-04,
derivative=-0.6829E-07
 iter num   12, nb calls=  24, f=  2.172
 linear search: initial derivative=-0.5587E-02
               step length= 1.0000    , df=-0.1244E-04,
derivative=-0.7963E-03
 iter num   13, nb calls=  25, f=  2.172
 linear search: initial derivative=-0.2601E-02
               step length= 0.9650    , df= 0.7277E-04,
derivative= 0.1780E-01
               step length= 0.1228    , df=-0.1583E-05,
derivative= 0.1707E-06
 iter num   14, nb calls=  27, f=  2.172
 linear search: initial derivative=-0.1302E-02
               step length= 0.3798    , df=-0.2898E-05,
derivative=-0.1081E-02
               step length=  2.228    , df=-0.9179E-05,
derivative= 0.3028E-04
 iter num   15, nb calls=  29, f=  2.172
 linear search: initial derivative=-0.1408E-02
               step length= 1.0000    , df=-0.1171E-05,
derivative=-0.1967E-03
 iter num   16, nb calls=  30, f=  2.172
 linear search: initial derivative=-0.1561E-03
               step length= 1.0000    , df=-0.2435E-07,
derivative= 0.1433E-05
 iter num   17, nb calls=  31, f=  2.172
 linear search: initial derivative=-0.2402E-05
```

```
              step length= 1.0000    , df=-0.1096E-10,
derivative= 0.1967E-06
 iter num    18, nb calls=  32, f=  2.172
 linear search: initial derivative=-0.3269E-06
              step length= 1.0000    , df=-0.1452E-12,
derivative= 0.5238E-08
 iter num    19, nb calls=  33, f=  2.172
 linear search: initial derivative=-0.1292E-07
              step length= 1.0000    , df=-0.8882E-15,
derivative= 0.1540E-08
 iter num    20, nb calls=  34, f=  2.172
 linear search: initial derivative=-0.2389E-08
              step length= 1.0000    , df=-0.8882E-15,
derivative= 0.1084E-09
 iter num    21, nb calls=  35, f=  2.172
 linear search: initial derivative=-0.1433E-09
              step length= 1.0000    , df= 0.0000E+00,
derivative= 0.1071E-09
              step length= 0.4644    , df= 0.8882E-15,
derivative= 0.1965E-10
              step length= 0.4644E-01, df=-0.4441E-15,
derivative=-0.1677E-09
              step length= 0.9288E-01, df= 0.8882E-15,
derivative=-0.6458E-10
              step length= 0.5108E-01, df= 0.0000E+00,
derivative=-0.8109E-10
              step length= 0.4691E-01, df= 0.0000E+00,
derivative=-0.3192E-10
              step length= 0.4649E-01, df= 0.4441E-15,
derivative=-0.1149E-09
              step length= 0.4645E-01, df= 0.8882E-15,
derivative=-0.1040E-09
              step length= 0.4644E-01, df=-0.4441E-15,
derivative=-0.1415E-09
              step length= 0.4644E-01, df= 0.0000E+00,
derivative=-0.1378E-09
              step length= 0.4644E-01, df=-0.8882E-15,
derivative=-0.9446E-10
 iter num    22, nb calls=  46, f=  2.172
 linear search: initial derivative=-0.9595E-10
              step length= 1.0000    , df= 0.0000E+00,
derivative= 0.4229E-09
              step length= 0.6270    , df= 0.8882E-15,
derivative= 0.1679E-09
              step length= 0.6270E-01, df= 0.8882E-15,
derivative=-0.5548E-10
              step length= 0.6270E-02, df= 0.1776E-14,
derivative= 0.1502E-10
              step length= 0.6270E-03, df= 0.2220E-14,
derivative=-0.7001E-10
```

```
                  step length= 0.6270E-04, df= 0.1332E-14,
derivative=-0.3073E-10
                  step length= 0.6270E-05, df= 0.1332E-14,
derivative=-0.1322E-09
                  step length= 0.6270E-06, df= 0.1776E-14,
derivative=-0.5178E-10
                  step length= 0.6270E-07, df= 0.0000E+00,
derivative=-0.9595E-10
                  step length= 0.1325E-07, df= 0.0000E+00,
derivative=-0.9595E-10
 iter num    22, nb calls= 56, f=  2.172
***** leaves -qn code-, gradient norm=  0.2468178351032304E-08
End of optimization.

  Iteration #4 calculation time =2.789 objective function =
2.1715974


***** enters -qn code- (without bound cstr)
dimension=        12, epsq=  0.2220446049250313E-15, verbosity
level: imp=        3
max number of iterations allowed: iter=         100
max number of calls to costf allowed: nap=        100
-------------------------------------------------
 iter num     1, nb calls=  1, f=  2.137
 linear search: initial derivative= -6.289
                  step length= 0.1000E-01, df=  9.814     ,
derivative=  51.10
                  step length= 0.1000E-02, df= 0.5142E-01,
derivative=  12.54
                  step length= 0.3607E-03, df=-0.2154E-01,
derivative=-0.1449E-01
 iter num     2, nb calls=  4, f=  2.115
 linear search: initial derivative= -1.042
                  step length= 0.1305E-01, df=  2.687     ,
derivative=  133.1
                  step length= 0.1305E-02, df= 0.2212E-01,
derivative=  11.76
                  step length= 0.1305E-03, df=-0.1675E-03,
derivative= 0.2319
 iter num     3, nb calls=  7, f=  2.115
 linear search: initial derivative=-0.4066
                  step length= 0.5802E-03, df=-0.2916E-03,
derivative=-0.3012
                  step length= 0.2239E-02, df=-0.6467E-03,
derivative=-0.4800E-03
 iter num     4, nb calls=  9, f=  2.115
 linear search: initial derivative=-0.3253
                  step length= 0.1487E-02, df=-0.1166E-02,
derivative=-0.2612
```

```
                step length= 0.7537E-02, df=-0.3257E-02,
derivative= 0.3851E-02
iter num      5, nb calls=  11, f=  2.111
linear search: initial derivative=-0.1236
                step length= 0.1234    , df= 0.1110E-01,
derivative= 0.5561
                step length= 0.2338E-01, df=-0.6192E-03,
derivative=-0.6083E-04
iter num      6, nb calls=  13, f=  2.111
linear search: initial derivative=-0.6284E-01
                step length= 0.7897E-01, df=-0.8388E-03,
derivative=-0.2235E-01
iter num      7, nb calls=  14, f=  2.110
linear search: initial derivative=-0.3045E-01
                step length= 0.4826    , df= 0.2631    ,
derivative=  9.457
                step length= 0.4826E-01, df= 0.2519E-02,
derivative= 0.9444
                step length= 0.4826E-02, df= 0.1010E-04,
derivative= 0.6713E-01
                step length= 0.1506E-02, df=-0.2617E-05,
derivative=-0.5461E-07
iter num      8, nb calls=  18, f=  2.110
linear search: initial derivative=-0.2677E-01
                step length= 0.6750E-03, df=-0.5218E-05,
derivative=-0.2660E-01
                step length= 0.6750E-02, df=-0.5064E-04,
derivative=-0.2503E-01
                step length= 0.6750E-01, df=-0.3506E-03,
derivative=-0.8944E-02
iter num      9, nb calls=  21, f=  2.110
linear search: initial derivative=-0.1782E-01
                step length= 0.4755    , df=-0.9837E-04,
derivative= 0.1245E-01
iter num     10, nb calls=  22, f=  2.109
linear search: initial derivative=-0.1983E-01
                step length= 0.5097    , df=-0.1493E-03,
derivative=-0.1006E-01
iter num     11, nb calls=  23, f=  2.109
linear search: initial derivative=-0.1064E-01
                step length= 0.6010    , df= 0.2716E-02,
derivative= 0.2049
                step length= 0.6010E-01, df= 0.1253E-06,
derivative= 0.1073E-01
                step length= 0.2993E-01, df=-0.7434E-05,
derivative=-0.4310E-07
iter num     12, nb calls=  26, f=  2.109
linear search: initial derivative=-0.7955E-02
```

```
                 step length= 0.9201E-01, df=-0.1383E-04,
derivative=-0.6841E-02
                 step length= 0.6533    , df=-0.5116E-04,
derivative= 0.4347E-03
 iter num    13, nb calls=  28, f=  2.109
 linear search: initial derivative=-0.5766E-02
                 step length= 0.5094    , df= 0.2310E-03,
derivative= 0.3196E-01
                 step length= 0.7875E-01, df=-0.7912E-05,
derivative= 0.1898E-05
 iter num    14, nb calls=  30, f=  2.109
 linear search: initial derivative=-0.2193E-02
                 step length= 1.0000    , df=-0.7246E-05,
derivative=-0.1217E-02
 iter num    15, nb calls=  31, f=  2.109
 linear search: initial derivative=-0.1033E-02
                 step length= 1.0000    , df=-0.4307E-05,
derivative= 0.1478E-04
 iter num    16, nb calls=  32, f=  2.109
 linear search: initial derivative=-0.7996E-03
                 step length= 1.0000    , df=-0.2647E-07,
derivative=-0.3822E-04
 iter num    17, nb calls=  33, f=  2.109
 linear search: initial derivative=-0.1237E-04
                 step length= 1.0000    , df=-0.2345E-09,
derivative= 0.3927E-06
 iter num    18, nb calls=  34, f=  2.109
 linear search: initial derivative=-0.1581E-05
                 step length= 1.0000    , df=-0.6106E-12,
derivative= 0.5513E-07
 iter num    19, nb calls=  35, f=  2.109
 linear search: initial derivative=-0.4802E-07
                 step length= 1.0000    , df=-0.8882E-15,
derivative= 0.6746E-09
 iter num    20, nb calls=  36, f=  2.109
 linear search: initial derivative=-0.2960E-08
                 step length= 1.0000    , df= 0.0000E+00,
derivative=-0.3086E-10
                 step length= 0.3316    , df= 0.8882E-15,
derivative=-0.1904E-08
                 step length= 0.3316E-01, df= 0.0000E+00,
derivative=-0.2868E-08
                 step length= 0.7095E-02, df= 0.0000E+00,
derivative=-0.2866E-08
                 step length= 0.1518E-02, df=-0.8882E-15,
derivative=-0.2852E-08
                 step length= 0.3037E-02, df= 0.0000E+00,
derivative=-0.2937E-08
                 step length= 0.1670E-02, df= 0.8882E-15,
derivative=-0.2843E-08
```

```
              step length= 0.1533E-02, df= 0.8882E-15,
derivative=-0.2861E-08
              step length= 0.1520E-02, df=-0.8882E-15,
derivative=-0.2872E-08
              step length= 0.1527E-02, df= 0.4441E-15,
derivative=-0.2906E-08
              step length= 0.1520E-02, df=-0.8882E-15,
derivative=-0.2908E-08
              step length= 0.1524E-02, df= 0.8882E-15,
derivative=-0.2915E-08
              step length= 0.1521E-02, df= 0.0000E+00,
derivative=-0.2804E-08
              step length= 0.1521E-02, df=-0.4441E-15,
derivative=-0.2898E-08
              step length= 0.1521E-02, df= 0.0000E+00,
derivative=-0.2796E-08
              step length= 0.1521E-02, df=-0.4441E-15,
derivative=-0.2898E-08
              step length= 0.1521E-02, df= 0.8882E-15,
derivative=-0.2844E-08
              step length= 0.1521E-02, df=-0.4441E-15,
derivative=-0.2898E-08
              step length= 0.1521E-02, df= 0.4441E-15,
derivative=-0.2829E-08
              step length= 0.1521E-02, df=-0.4441E-15,
derivative=-0.2898E-08
              step length= 0.1521E-02, df= 0.0000E+00,
derivative=-0.2843E-08
              step length= 0.1521E-02, df=-0.4441E-15,
derivative=-0.2898E-08
              step length= 0.1521E-02, df= 0.0000E+00,
derivative=-0.2827E-08
 iter num   21, nb calls=  59, f=  2.109
 linear search: initial derivative=-0.3594E-08
              step length= 1.0000     , df= 0.0000E+00,
derivative= 0.1397E-09
              step length= 0.3399     , df= 0.8882E-15,
derivative=-0.2441E-08
              step length= 0.3399E-01, df= 0.1776E-14,
derivative=-0.3518E-08
              step length= 0.3399E-02, df= 0.0000E+00,
derivative=-0.3531E-08
              step length= 0.7232E-03, df= 0.0000E+00,
derivative=-0.3625E-08
              step length= 0.1523E-03, df= 0.0000E+00,
derivative=-0.3622E-08
              step length= 0.3209E-04, df= 0.3109E-14,
derivative=-0.3557E-08
              step length= 0.3209E-05, df= 0.0000E+00,
derivative=-0.3547E-08
```

```
                step length= 0.6816E-06, df= 0.1776E-14,
derivative=-0.3453E-08
                step length= 0.6816E-07, df= 0.0000E+00,
derivative=-0.3552E-08
                step length= 0.1447E-07, df= 0.0000E+00,
derivative=-0.3594E-08
                step length= 0.3058E-08, df= 0.0000E+00,
derivative=-0.3594E-08
 iter num    22, nb calls=  71, f=  2.109
 linear search: initial derivative=-0.3201E-09
                step length= 1.0000     , df=-0.4441E-15,
derivative= 0.1179E-09
 iter num    23, nb calls=  72, f=  2.109
 linear search: initial derivative=-0.1202E-09
                step length= 1.0000     , df= 0.1332E-14,
derivative= 0.2276E-11
                step length= 0.1000E+00, df= 0.1776E-14,
derivative= 0.4688E-10
                step length= 0.1000E-01, df= 0.2220E-14,
derivative=-0.4078E-10
                step length= 0.1000E-02, df= 0.4441E-15,
derivative=-0.1300E-10
                step length= 0.1000E-03, df= 0.8882E-15,
derivative=-0.4180E-10
                step length= 0.1000E-04, df=-0.4441E-15,
derivative=-0.8874E-10
                step length= 0.2000E-04, df= 0.1332E-14,
derivative=-0.1406E-09
                step length= 0.1100E-04, df= 0.4441E-15,
derivative=-0.1801E-09
                step length= 0.1010E-04, df=-0.4441E-15,
derivative=-0.8874E-10
                step length= 0.1055E-04, df=-0.4441E-15,
derivative=-0.8874E-10
                step length= 0.1078E-04, df= 0.1332E-14,
derivative=-0.1688E-09
                step length= 0.1057E-04, df= 0.4441E-15,
derivative=-0.9432E-10
 iter num    24, nb calls=  84, f=  2.109
 linear search: initial derivative=-0.8819E-10
                step length= 1.0000     , df= 0.1776E-14,
derivative= 0.5672E-10
                step length= 0.1000E+00, df= 0.2665E-14,
derivative= 0.7731E-10
                step length= 0.1000E-01, df= 0.2220E-14,
derivative=-0.6815E-10
                step length= 0.1000E-02, df= 0.2665E-14,
derivative=-0.3128E-10
                step length= 0.1000E-03, df= 0.2665E-14,
derivative=-0.6813E-10
```

```
              step length= 0.1000E-04, df= 0.2665E-14,
derivative=-0.3935E-10
              step length= 0.1000E-05, df= 0.8882E-15,
derivative=-0.5833E-11
              step length= 0.1000E-06, df= 0.0000E+00,
derivative=-0.8819E-10
              step length= 0.2113E-07, df= 0.0000E+00,
derivative=-0.8819E-10
  iter num   24, nb calls=  93, f=  2.109
***** leaves -qn code-, gradient norm=  0.5847386598097223E-09
End of optimization.

  Iteration #5 calculation time =4.731 objective function =
2.1092132


***** enters -qn code- (without bound cstr)
dimension=        12, epsq=  0.2220446049250313E-15, verbosity
level: imp=        3
max number of iterations allowed: iter=         100
max number of calls to costf allowed: nap=        100
--------------------------------------------------
  iter num    1, nb calls=   1, f=  2.085
  linear search: initial derivative= -5.108
              step length= 0.1000E-01, df=  9.577      ,
derivative=  51.33
              step length= 0.1000E-02, df= 0.7316E-01,
derivative=  13.52
              step length= 0.2925E-03, df=-0.1412E-01,
derivative=-0.1886E-01
  iter num    2, nb calls=   4, f=  2.071
  linear search: initial derivative=-0.8632
              step length= 0.1010E-01, df=  1.532      ,
derivative=  95.72
              step length= 0.1010E-02, df= 0.1242E-01,
derivative=  8.469
              step length= 0.1010E-03, df=-0.1303E-03,
derivative= 0.6649E-01
  iter num    3, nb calls=   7, f=  2.071
  linear search: initial derivative=-0.3719
              step length= 0.4213E-03, df=-0.2304E-03,
derivative=-0.2855
              step length= 0.1814E-02, df=-0.5614E-03,
derivative=-0.4892E-03
  iter num    4, nb calls=   9, f=  2.071
  linear search: initial derivative=-0.2994
              step length= 0.1210E-02, df=-0.1002E-02,
derivative=-0.2348
              step length= 0.5600E-02, df=-0.2583E-02,
derivative= 0.3527E-02
  iter num    5, nb calls=  11, f=  2.068
```

```
linear search: initial derivative=-0.9516E-01
                step length= 0.1408     , df= 0.8863E-02,
derivative= 0.4212
                step length= 0.2590E-01, df=-0.4751E-03,
derivative=-0.1143E-04
 iter num    6, nb calls=   13, f=   2.067
 linear search: initial derivative=-0.6490E-01
                step length= 0.3317E-01, df=-0.7254E-03,
derivative=-0.3414E-01
 iter num    7, nb calls=   14, f=   2.067
 linear search: initial derivative=-0.2984E-01
                step length= 0.3889     , df= 0.3093E-01,
derivative= 1.273
                step length= 0.3889E-01, df= 0.1912E-03,
derivative= 0.1082
                step length= 0.8367E-02, df=-0.1560E-04,
derivative=-0.4938E-07
 iter num    8, nb calls=   17, f=   2.067
 linear search: initial derivative=-0.2938E-01
                step length= 0.3996E-03, df=-0.3073E-04,
derivative=-0.2850E-01
                step length= 0.3996E-02, df=-0.2651E-03,
derivative=-0.2058E-01
                step length= 0.1338E-01, df=-0.5249E-03,
derivative=-0.3675E-03
 iter num    9, nb calls=   20, f=   2.066
 linear search: initial derivative=-0.1997E-01
                step length= 0.5826     , df= 0.1127E-02,
derivative= 0.6629E-01
                step length= 0.1481     , df=-0.1344E-03,
derivative= 0.5448E-04
 iter num   10, nb calls=   22, f=   2.066
 linear search: initial derivative=-0.1435E-01
                step length= 0.5107     , df=-0.1792E-03,
derivative=-0.4698E-02
 iter num   11, nb calls=   23, f=   2.066
 linear search: initial derivative=-0.4705E-02
                step length= 1.0000     , df= 0.9427E-03,
derivative= 0.8529E-01
                step length= 0.1000E+00, df=-0.5317E-06,
derivative= 0.4253E-02
                step length= 0.5254E-01, df=-0.2898E-05,
derivative= 0.6949E-07
 iter num   12, nb calls=   26, f=   2.066
 linear search: initial derivative=-0.4130E-02
                step length= 0.2453E-01, df=-0.5605E-05,
derivative=-0.3859E-02
                step length= 0.2453     , df=-0.3878E-04,
derivative=-0.1385E-02
 iter num   13, nb calls=   28, f=   2.066
```

```
linear search: initial derivative=-0.1797E-02
            step length= 1.0000    , df= 0.4445E-04,
derivative= 0.9071E-02
            step length= 0.1658    , df=-0.1824E-05,
derivative= 0.2751E-06
iter num    14, nb calls=  30, f=  2.066
linear search: initial derivative=-0.1565E-02
            step length= 0.8995E-01, df=-0.3179E-05,
derivative=-0.1163E-02
            step length= 0.3493    , df=-0.7055E-05,
derivative= 0.1302E-04
iter num    15, nb calls=  32, f=  2.066
linear search: initial derivative=-0.1386E-02
            step length= 1.0000    , df=-0.3085E-06,
derivative=-0.2813E-03
iter num    16, nb calls=  33, f=  2.066
linear search: initial derivative=-0.2303E-03
            step length= 1.0000    , df=-0.1515E-07,
derivative=-0.4048E-06
iter num    17, nb calls=  34, f=  2.066
linear search: initial derivative=-0.2909E-05
            step length= 1.0000    , df=-0.1126E-10,
derivative= 0.1110E-06
iter num    18, nb calls=  35, f=  2.066
linear search: initial derivative=-0.1467E-05
            step length= 1.0000    , df=-0.2620E-12,
derivative= 0.9629E-07
iter num    19, nb calls=  36, f=  2.066
linear search: initial derivative=-0.4714E-07
            step length= 1.0000    , df=-0.3553E-14,
derivative= 0.3037E-09
iter num    20, nb calls=  37, f=  2.066
linear search: initial derivative=-0.3680E-09
            step length= 1.0000    , df=-0.4441E-15,
derivative=-0.8965E-11
iter num    21, nb calls=  38, f=  2.066
linear search: initial derivative=-0.7851E-10
            step length= 1.0000    , df= 0.0000E+00,
derivative= 0.7847E-10
            step length= 0.4999    , df=-0.8882E-15,
derivative= 0.5479E-10
iter num    22, nb calls=  40, f=  2.066
linear search: initial derivative=-0.1301E-09
            step length= 1.0000    , df= 0.0000E+00,
derivative= 0.1968E-09
            step length= 0.5495    , df= 0.1776E-14,
derivative= 0.4805E-10
            step length= 0.5495E-01, df= 0.1332E-14,
derivative=-0.3335E-10
```

```
                step length= 0.5495E-02, df= 0.1332E-14,
derivative=-0.9264E-10
                step length= 0.5495E-03, df= 0.8882E-15,
derivative=-0.1156E-10
                step length= 0.5495E-04, df= 0.0000E+00,
derivative=-0.2142E-09
                step length= 0.9218E-05, df= 0.8882E-15,
derivative=-0.3434E-10
                step length= 0.9218E-06, df= 0.8882E-15,
derivative=-0.1038E-09
                step length= 0.9218E-07, df= 0.0000E+00,
derivative=-0.1301E-09
 iter num    23, nb calls= 49, f=  2.066
 linear search: initial derivative=-0.2609E-09
                step length= 1.0000     , df= 0.8882E-15,
derivative= 0.3730E-09
                step length= 0.1000E+00, df= 0.1332E-14,
derivative=-0.2039E-09
                step length= 0.1000E-01, df= 0.8882E-15,
derivative=-0.2911E-09
                step length= 0.1000E-02, df= 0.1776E-14,
derivative=-0.2602E-09
                step length= 0.1000E-03, df= 0.8882E-15,
derivative=-0.2737E-09
                step length= 0.1000E-04, df=-0.8882E-15,
derivative=-0.2799E-09
                step length= 0.2000E-04, df= 0.8882E-15,
derivative=-0.4257E-09
                step length= 0.1100E-04, df= 0.8882E-15,
derivative=-0.2577E-09
                step length= 0.1010E-04, df=-0.8882E-15,
derivative=-0.2799E-09
                step length= 0.1055E-04, df= 0.8882E-15,
derivative=-0.3749E-09
                step length= 0.1015E-04, df=-0.8882E-15,
derivative=-0.2799E-09
                step length= 0.1035E-04, df= 0.8882E-15,
derivative=-0.3749E-09
                step length= 0.1017E-04, df=-0.8882E-15,
derivative=-0.2799E-09
                step length= 0.1026E-04, df=-0.8882E-15,
derivative=-0.2799E-09
                step length= 0.1030E-04, df= 0.8882E-15,
derivative=-0.3749E-09
 iter num    24, nb calls= 64, f=  2.066
 linear search: initial derivative=-0.4359E-09
                step length= 1.0000     , df= 0.8882E-15,
derivative= 0.5039E-09
                step length= 0.1000E+00, df= 0.4441E-15,
derivative=-0.2641E-09
```

```
                step length= 0.1000E-01, df= 0.1332E-14,
derivative=-0.4545E-09
                step length= 0.1000E-02, df= 0.2665E-14,
derivative=-0.4414E-09
                step length= 0.1000E-03, df= 0.1776E-14,
derivative=-0.3762E-09
                step length= 0.1000E-04, df= 0.1776E-14,
derivative=-0.4088E-09
                step length= 0.1000E-05, df= 0.8882E-15,
derivative=-0.3045E-09
                step length= 0.1000E-06, df= 0.0000E+00,
derivative=-0.4359E-09
  iter num    24, nb calls=  72, f=  2.066
***** leaves -qn code-, gradient norm=  0.7803205599769735E-09
End of optimization.
  Iteration #6 calculation time =4.19 objective function =
2.065834

***** enters -qn code- (without bound cstr)
dimension=        12, epsq=  0.2220446049250313E-15, verbosity
level: imp=        3
max number of iterations allowed: iter=        100
max number of calls to costf allowed: nap=        100
--------------------------------------------------
  iter num     1, nb calls=   1, f=  2.049
  linear search: initial derivative= -4.190
                step length= 0.1000E-01, df=  9.303     ,
derivative=  51.13
                step length= 0.1000E-02, df=  0.8623E-01,
derivative=  13.89
                step length= 0.2446E-03, df=-0.9641E-02,
derivative=-0.1937E-01
  iter num     2, nb calls=   4, f=  2.039
  linear search: initial derivative=-0.7493
                step length= 0.7496E-02, df=  0.8982     ,
derivative=  71.23
                step length= 0.7496E-03, df=  0.7083E-02,
derivative=  6.262
                step length= 0.8035E-04, df=-0.1033E-03,
derivative=-0.5825E-05
  iter num     3, nb calls=   7, f=  2.039
  linear search: initial derivative=-0.3323
                step length= 0.3391E-03, df=-0.1825E-03,
derivative=-0.2545
                step length= 0.1449E-02, df=-0.4419E-03,
derivative=-0.3933E-03
  iter num     4, nb calls=   9, f=  2.039
  linear search: initial derivative=-0.2666
                step length= 0.9999E-03, df=-0.7845E-03,
derivative=-0.2067
```

```
              step length= 0.4443E-02, df=-0.1955E-02,
derivative= 0.2362E-02
 iter num     5, nb calls=  11, f=  2.037
 linear search: initial derivative=-0.8501E-01
              step length= 0.1096    , df= 0.1994E-02,
derivative= 0.1724
              step length= 0.3641E-01, df=-0.6494E-03,
derivative= 0.1960E-04
 iter num     6, nb calls=  13, f=  2.036
 linear search: initial derivative=-0.6533E-01
              step length= 0.2813E-01, df=-0.5764E-03,
derivative= 0.7627E-02
 iter num     7, nb calls=  14, f=  2.035
 linear search: initial derivative=-0.3985E-01
              step length= 0.1051    , df= 0.1998E-01,
derivative=  1.383
              step length= 0.1051E-01, df= 0.1061E-03,
derivative= 0.1128
              step length= 0.2728E-02, df=-0.1495E-04,
derivative= 0.4361E-07
 iter num     8, nb calls=  17, f=  2.035
 linear search: initial derivative=-0.3830E-01
              step length= 0.3606E-03, df=-0.2934E-04,
derivative=-0.3684E-01
              step length= 0.3606E-02, df=-0.2429E-03,
derivative=-0.2398E-01
 iter num     9, nb calls=  19, f=  2.035
 linear search: initial derivative=-0.2054E-01
              step length= 0.2086    , df=-0.2552E-03,
derivative=-0.1118E-02
 iter num    10, nb calls=  20, f=  2.035
 linear search: initial derivative=-0.1338E-01
              step length= 0.6853    , df= 0.4545E-03,
derivative= 0.3722E-01
              step length= 0.1812    , df=-0.6743E-04,
derivative= 0.6534E-05
 iter num    11, nb calls=  22, f=  2.035
 linear search: initial derivative=-0.1309E-01
              step length= 0.1749E-01, df=-0.1131E-03,
derivative=-0.8866E-02
 iter num    12, nb calls=  23, f=  2.035
 linear search: initial derivative=-0.8244E-02
              step length= 0.4685    , df= 0.5042E-03,
derivative= 0.4503E-01
              step length= 0.7255E-01, df=-0.1751E-04,
derivative= 0.1302E-05
 iter num    13, nb calls=  25, f=  2.035
 linear search: initial derivative=-0.7661E-02
              step length= 0.1154    , df=-0.3167E-04,
derivative=-0.6194E-02
```

```
                step length= 0.6023    , df=-0.9091E-04,
derivative= 0.7775E-04
 iter num    14, nb calls=  27, f=  2.035
 linear search: initial derivative=-0.2313E-02
                step length= 1.0000    , df=-0.8711E-06,
derivative=-0.1308E-03
 iter num    15, nb calls=  28, f=  2.035
 linear search: initial derivative=-0.1282E-03
                step length= 1.0000    , df=-0.3363E-07,
derivative=-0.4515E-05
 iter num    16, nb calls=  29, f=  2.035
 linear search: initial derivative=-0.9779E-05
                step length= 1.0000    , df=-0.1015E-09,
derivative=-0.3408E-06
 iter num    17, nb calls=  30, f=  2.035
 linear search: initial derivative=-0.1679E-05
                step length= 1.0000    , df=-0.3491E-12,
derivative=-0.3851E-07
 iter num    18, nb calls=  31, f=  2.035
 linear search: initial derivative=-0.7391E-07
                step length= 1.0000    , df= 0.8882E-15,
derivative= 0.5200E-09
                step length= 0.1690    , df= 0.4441E-15,
derivative=-0.6135E-07
                step length= 0.1690E-01, df= 0.4441E-15,
derivative=-0.7263E-07
                step length= 0.1690E-02, df=-0.1332E-14,
derivative=-0.7394E-07
                step length= 0.3380E-02, df= 0.4441E-15,
derivative=-0.7372E-07
                step length= 0.1859E-02, df= 0.8882E-15,
derivative=-0.7382E-07
                step length= 0.1707E-02, df= 0.0000E+00,
derivative=-0.7378E-07
                step length= 0.1692E-02, df= 0.0000E+00,
derivative=-0.7389E-07
                step length= 0.1690E-02, df= 0.1332E-14,
derivative=-0.7381E-07
                step length= 0.1690E-02, df= 0.0000E+00,
derivative=-0.7393E-07
                step length= 0.1690E-02, df= 0.0000E+00,
derivative=-0.7387E-07
 iter num    19, nb calls=  42, f=  2.035
 linear search: initial derivative=-0.7272E-07
                step length= 1.0000    , df= 0.1776E-14,
derivative= 0.5857E-09
                step length= 0.1131    , df= 0.8882E-15,
derivative=-0.6431E-07
                step length= 0.1131E-01, df= 0.1332E-14,
derivative=-0.7177E-07
```

```
                step length= 0.1131E-02, df= 0.4441E-15,
derivative=-0.7258E-07
                step length= 0.1131E-03, df= 0.2220E-14,
derivative=-0.7251E-07
                step length= 0.1131E-04, df= 0.1332E-14,
derivative=-0.7271E-07
                step length= 0.1131E-05, df= 0.4441E-15,
derivative=-0.7261E-07
                step length= 0.1131E-06, df= 0.1332E-14,
derivative=-0.7254E-07
                step length= 0.1131E-07, df= 0.4441E-15,
derivative=-0.7269E-07
                step length= 0.1131E-08, df= 0.0000E+00,
derivative=-0.7272E-07
 iter num     19, nb calls=  52, f=  2.035
***** leaves -qn code-, gradient norm=  0.1044663447000345E-05
End of optimization.

   Iteration #7 calculation time =2.861 objective function =
2.0345853

***** enters -qn code- (without bound cstr)
dimension=          12, epsq=  0.2220446049250313E-15, verbosity
level: imp=          3
max number of iterations allowed: iter=          100
max number of calls to costf allowed: nap=        100
------------------------------------------------
 iter num      1, nb calls=   1, f=  2.022
 linear search: initial derivative= -3.475
                step length= 0.1000E-01, df=  9.022      ,
derivative=  50.73
                step length= 0.1000E-02, df= 0.9390E-01,
derivative=  13.93
                step length= 0.2089E-03, df=-0.6801E-02,
derivative=-0.1807E-01
 iter num      2, nb calls=   4, f=  2.015
 linear search: initial derivative=-0.6663
                step length= 0.5533E-02, df= 0.5458      ,
derivative=  54.55
                step length= 0.5533E-03, df= 0.4155E-02,
derivative=  4.742
                step length= 0.6832E-04, df=-0.8399E-04,
derivative=-0.3204E-05
 iter num      3, nb calls=   7, f=  2.015
 linear search: initial derivative=-0.2949
                step length= 0.2913E-03, df=-0.1469E-03,
derivative=-0.2211
```

```
                step length= 0.1164E-02, df=-0.3357E-03,
derivative=-0.2806E-03
 iter num     4, nb calls=    9, f=   2.015
 linear search: initial derivative=-0.2351
                step length= 0.8290E-03, df=-0.5979E-03,
derivative=-0.1836
                step length= 0.3777E-02, df=-0.1524E-02,
derivative= 0.1650E-02
 iter num     5, nb calls=   11, f=   2.013
 linear search: initial derivative=-0.8402E-01
                step length= 0.7065E-01, df= 0.2906E-03,
derivative= 0.1046
                step length= 0.3274E-01, df=-0.7141E-03,
derivative= 0.8223E-05
 iter num     6, nb calls=   13, f=   2.013
 linear search: initial derivative=-0.5451E-01
                step length= 0.5221E-01, df= 0.1122E-03,
derivative= 0.6260E-01
                step length= 0.2414E-01, df=-0.3296E-03,
derivative= 0.7938E-06
 iter num     7, nb calls=   15, f=   2.012
 linear search: initial derivative=-0.4435E-01
                step length= 0.2126E-01, df= 0.2820E-02,
derivative= 0.4169
                step length= 0.2126E-02, df=-0.3018E-04,
derivative= 0.3666E-02
 iter num     8, nb calls=   17, f=   2.012
 linear search: initial derivative=-0.4318E-01
                step length= 0.3164E-03, df=-0.5818E-04,
derivative=-0.4006E-01
                step length= 0.3164E-02, df=-0.3883E-03,
derivative=-0.1260E-01
 iter num     9, nb calls=   19, f=   2.012
 linear search: initial derivative=-0.1631E-01
                step length= 0.4817    , df= 0.7015E-04,
derivative= 0.1901E-01
                step length= 0.2200    , df=-0.1766E-03,
derivative= 0.3331E-05
 iter num    10, nb calls=   21, f=   2.012
 linear search: initial derivative=-0.9806E-02
                step length= 0.5076    , df= 0.5768E-03,
derivative= 0.4182E-01
                step length= 0.9632E-01, df=-0.3347E-04,
derivative= 0.5926E-05
 iter num    11, nb calls=   23, f=   2.012
 linear search: initial derivative=-0.9430E-02
                step length= 0.1300E-01, df=-0.5996E-04,
derivative=-0.7462E-02
```

```
                step length= 0.6241E-01, df=-0.1610E-03,
derivative=-0.1833E-04
 iter num    12, nb calls=  25, f=  2.011
 linear search: initial derivative=-0.6499E-02
                step length= 0.7630    , df= 0.3035E-02,
derivative= 0.1291
                step length= 0.7630E-01, df= 0.1349E-05,
derivative= 0.7044E-02
                step length= 0.3662E-01, df=-0.7725E-05,
derivative= 0.3382E-07
 iter num    13, nb calls=  28, f=  2.011
 linear search: initial derivative=-0.5321E-02
                step length= 1.0000    , df=-0.1299E-05,
derivative= 0.2186E-03
 iter num    14, nb calls=  29, f=  2.011
 linear search: initial derivative=-0.1635E-03
                step length= 1.0000    , df=-0.1713E-07,
derivative=-0.8146E-05
 iter num    15, nb calls=  30, f=  2.011
 linear search: initial derivative=-0.1268E-04
                step length= 1.0000    , df=-0.1962E-09,
derivative=-0.7951E-06
 iter num    16, nb calls=  31, f=  2.011
 linear search: initial derivative=-0.3469E-05
                step length= 1.0000    , df=-0.2318E-11,
derivative=-0.2932E-07
 iter num    17, nb calls=  32, f=  2.011
 linear search: initial derivative=-0.1970E-06
                step length= 1.0000    , df=-0.1865E-13,
derivative= 0.2056E-07
 iter num    18, nb calls=  33, f=  2.011
 linear search: initial derivative=-0.1435E-07
                step length= 1.0000    , df= 0.0000E+00,
derivative= 0.2467E-09
                step length= 0.3362    , df= 0.0000E+00,
derivative=-0.9423E-08
                step length= 0.8198E-01, df=-0.1332E-14,
derivative=-0.1308E-07
                step length= 0.1640    , df=-0.1332E-14,
derivative=-0.1197E-07
                step length= 0.2501    , df= 0.4441E-15,
derivative=-0.1057E-07
                step length= 0.1726    , df= 0.4441E-15,
derivative=-0.1182E-07
                step length= 0.1648    , df=-0.8882E-15,
derivative=-0.1188E-07
                step length= 0.1687    , df= 0.8882E-15,
derivative=-0.1189E-07
                step length= 0.1652    , df=-0.8882E-15,
derivative=-0.1186E-07
```

```
                step length= 0.1669    , df=-0.4441E-15,
derivative=-0.1189E-07
                step length= 0.1678    , df= 0.4441E-15,
derivative=-0.1180E-07
                step length= 0.1670    , df= 0.8882E-15,
derivative=-0.1187E-07
                step length= 0.1670    , df= 0.1332E-14,
derivative=-0.1182E-07
                step length= 0.1669    , df=-0.8882E-15,
derivative=-0.1193E-07
                step length= 0.1670    , df= 0.0000E+00,
derivative=-0.1185E-07
                step length= 0.1669    , df=-0.4441E-15,
derivative=-0.1186E-07
                step length= 0.1670    , df= 0.4441E-15,
derivative=-0.1183E-07
                step length= 0.1669    , df= 0.4441E-15,
derivative=-0.1195E-07
                step length= 0.1669    , df= 0.4441E-15,
derivative=-0.1197E-07
                step length= 0.1669    , df= 0.0000E+00,
derivative=-0.1191E-07
                step length= 0.1669    , df= 0.0000E+00,
derivative=-0.1191E-07
 iter num   19, nb calls=  54, f=  2.011
 linear search: initial derivative=-0.1200E-07
                step length= 1.0000    , df= 0.1776E-14,
derivative= 0.1144E-09
                step length= 0.1000E+00, df= 0.8882E-15,
derivative=-0.1075E-07
                step length= 0.1000E-01, df= 0.2665E-14,
derivative=-0.1177E-07
                step length= 0.1000E-02, df= 0.8882E-15,
derivative=-0.1184E-07
                step length= 0.1000E-03, df= 0.1776E-14,
derivative=-0.1184E-07
                step length= 0.1000E-04, df= 0.8882E-15,
derivative=-0.1194E-07
                step length= 0.1000E-05, df= 0.1776E-14,
derivative=-0.1198E-07
                step length= 0.1000E-06, df= 0.8882E-15,
derivative=-0.1192E-07
                step length= 0.1000E-07, df= 0.8882E-15,
derivative=-0.1194E-07
                step length= 0.1000E-08, df= 0.0000E+00,
derivative=-0.1200E-07
 iter num   19, nb calls=  64, f=  2.011
***** leaves -qn code-, gradient norm=  0.6239573108403149E-06
End of optimization.
```

Iteration #8 calculation time =3.546 objective function =
2.0114014

***** enters -qn code- (without bound cstr)
dimension=   12, epsq= 0.2220446049250313E-15, verbosity
level: imp=   3
max number of iterations allowed: iter=   100
max number of calls to costf allowed: nap=   100
----
 iter num   1, nb calls=   1, f= 2.002
 linear search: initial derivative= -2.913
                step length= 0.1000E-01, df= 8.755   ,
derivative= 50.30
                step length= 0.1000E-02, df= 0.9819E-01,
derivative= 13.79
                step length= 0.1812E-03, df=-0.4927E-02,
derivative=-0.1612E-01
 iter num   2, nb calls=   4, f= 1.997
 linear search: initial derivative=-0.5995
                step length= 0.4139E-02, df= 0.3440   ,
derivative= 42.72
                step length= 0.4139E-03, df= 0.2513E-02,
derivative= 3.660
                step length= 0.5835E-04, df=-0.6948E-04,
derivative=-0.1843E-05
 iter num   3, nb calls=   7, f= 1.997
 linear search: initial derivative=-0.2622
                step length= 0.2577E-03, df=-0.1201E-03,
derivative=-0.1910
                step length= 0.9493E-03, df=-0.2560E-03,
derivative=-0.1849E-03
 iter num   4, nb calls=   9, f= 1.997
 linear search: initial derivative=-0.2082
                step length= 0.6872E-03, df=-0.4600E-03,
derivative=-0.1658
                step length= 0.3371E-02, df=-0.1252E-02,
derivative= 0.1380E-02
 iter num   5, nb calls=   11, f= 1.995
 linear search: initial derivative=-0.8253E-01
                step length= 0.4883E-01, df= 0.6996E-03,
derivative= 0.1347
                step length= 0.1953E-01, df=-0.5061E-03,
derivative= 0.2305E-04
 iter num   6, nb calls=   13, f= 1.995
 linear search: initial derivative=-0.5044E-01
                step length= 0.3931E-01, df=-0.6557E-03,
derivative=-0.1507E-01
 iter num   7, nb calls=   14, f= 1.994
 linear search: initial derivative=-0.4206E-01

```
                    step length= 0.3441E-01, df= 0.5405E-01,
derivative=  3.382
                    step length= 0.3441E-02, df= 0.4585E-03,
derivative= 0.3349
                    step length= 0.3804E-03, df=-0.7246E-05,
derivative=-0.3743E-07
 iter num     8, nb calls=  17, f=  1.994
 linear search: initial derivative=-0.3467E-01
                    step length= 0.5946E-03, df=-0.1373E-04,
derivative=-0.3102E-01
                    step length= 0.5650E-02, df=-0.6895E-04,
derivative=-0.8018E-04
 iter num     9, nb calls=  19, f=  1.994
 linear search: initial derivative=-0.1381E-01
                    step length= 0.1051    , df=-0.1120E-03,
derivative=-0.8622E-02
 iter num    10, nb calls=  20, f=  1.994
 linear search: initial derivative=-0.1005E-01
                    step length= 0.1965    , df=-0.1608E-03,
derivative=-0.4401E-02
 iter num    11, nb calls=  21, f=  1.994
 linear search: initial derivative=-0.7844E-02
                    step length= 0.3080    , df= 0.4897E-02,
derivative= 0.2465
                    step length= 0.3080E-01, df= 0.2011E-04,
derivative= 0.1765E-01
                    step length= 0.9474E-02, df=-0.4948E-05,
derivative= 0.2396E-08
 iter num    12, nb calls=  24, f=  1.994
 linear search: initial derivative=-0.6642E-02
                    step length= 0.5146E-01, df=-0.9267E-05,
derivative=-0.5797E-02
                    step length= 0.4044    , df=-0.3884E-04,
derivative= 0.1262E-04
 iter num    13, nb calls=  26, f=  1.994
 linear search: initial derivative=-0.6467E-02
                    step length= 0.6333E-01, df=-0.3479E-04,
derivative= 0.6853E-03
 iter num    14, nb calls=  27, f=  1.994
 linear search: initial derivative=-0.4760E-02
                    step length= 1.0000    , df=-0.1980E-05,
derivative=-0.2102E-04
 iter num    15, nb calls=  28, f=  1.994
 linear search: initial derivative=-0.1568E-03
                    step length= 1.0000    , df=-0.3146E-07,
derivative=-0.6005E-05
 iter num    16, nb calls=  29, f=  1.994
 linear search: initial derivative=-0.3043E-04
```

```
              step length= 1.0000    , df=-0.4014E-09,
derivative= 0.1098E-05
 iter num   17, nb calls= 30, f=  1.994
 linear search: initial derivative=-0.2455E-05
              step length= 1.0000    , df=-0.2088E-11,
derivative= 0.1123E-06
 iter num   18, nb calls= 31, f=  1.994
 linear search: initial derivative=-0.1190E-06
              step length= 1.0000    , df=-0.4219E-14,
derivative= 0.1571E-09
 iter num   19, nb calls= 32, f=  1.994
 linear search: initial derivative=-0.1059E-07
              step length= 1.0000    , df=-0.1110E-14,
derivative=-0.1638E-09
 iter num   20, nb calls= 33, f=  1.994
 linear search: initial derivative=-0.2251E-09
              step length= 1.0000    , df= 0.1110E-14,
derivative= 0.1520E-09
              step length= 0.1000E+00, df= 0.6661E-15,
derivative=-0.1069E-09
              step length= 0.1000E-01, df=-0.2220E-15,
derivative=-0.2302E-09
              step length= 0.2000E-01, df= 0.4441E-15,
derivative=-0.1401E-09
              step length= 0.1100E-01, df= 0.8882E-15,
derivative=-0.1490E-09
              step length= 0.1010E-01, df=-0.4441E-15,
derivative=-0.6141E-10
 iter num   21, nb calls= 39, f=  1.994
 linear search: initial derivative=-0.1326E-08
              step length= 1.0000    , df= 0.0000E+00,
derivative= 0.1960E-08
              step length= 0.5470    , df= 0.1110E-14,
derivative= 0.4655E-09
              step length= 0.5470E-01, df= 0.1554E-14,
derivative=-0.1104E-08
              step length= 0.5470E-02, df= 0.1776E-14,
derivative=-0.1311E-08
              step length= 0.5470E-03, df= 0.8882E-15,
derivative=-0.1288E-08
              step length= 0.5470E-04, df= 0.0000E+00,
derivative=-0.1280E-08
              step length= 0.1172E-04, df= 0.6661E-15,
derivative=-0.1154E-08
              step length= 0.1172E-05, df= 0.0000E+00,
derivative=-0.1326E-08
              step length= 0.2477E-06, df= 0.0000E+00,
derivative=-0.1326E-08
 iter num   22, nb calls= 48, f=  1.994
```

```
linear search: initial derivative=-0.9526E-09
               step length= 1.0000    , df= 0.0000E+00,
derivative= 0.1281E-08
               step length= 0.5362    , df= 0.8882E-15,
derivative= 0.2651E-09
               step length= 0.5362E-01, df= 0.2220E-15,
derivative=-0.8799E-09
               step length= 0.5362E-02, df= 0.4441E-15,
derivative=-0.9555E-09
               step length= 0.5362E-03, df= 0.1554E-14,
derivative=-0.9727E-09
               step length= 0.5362E-04, df= 0.4441E-15,
derivative=-0.8848E-09
               step length= 0.5362E-05, df= 0.1332E-14,
derivative=-0.8703E-09
               step length= 0.5362E-06, df= 0.0000E+00,
derivative=-0.9526E-09
               step length= 0.1133E-06, df= 0.0000E+00,
derivative=-0.9526E-09
 iter num   22, nb calls=  57, f=  1.994
***** leaves -qn code-, gradient norm=  0.6916385691667439E-08
End of optimization.

   Iteration #9 calculation time =3.819 objective function =
1.9937792

***** enters -qn code- (without bound cstr)
dimension=        12, epsq=  0.2220446049250313E-15, verbosity
level: imp=        3
max number of iterations allowed: iter=        100
max number of calls to costf allowed: nap=        100
-----------------------------------------------
 iter num    1, nb calls=   1, f=  1.986
 linear search: initial derivative= -2.468
               step length= 0.1000E-01, df=  8.510     ,
derivative=  49.89
               step length= 0.1000E-02, df= 0.1004     ,
derivative=  13.57
               step length= 0.1591E-03, df=-0.3652E-02,
derivative=-0.1410E-01
 iter num    2, nb calls=   4, f=  1.983
 linear search: initial derivative=-0.5429
               step length= 0.3157E-02, df= 0.2243     ,
derivative=  34.06
               step length= 0.3157E-03, df= 0.1564E-02,
derivative=  2.870
               step length= 0.5028E-04, df=-0.5817E-04,
derivative=-0.1100E-05
 iter num    3, nb calls=   7, f=  1.983
```

```
linear search: initial derivative=-0.2351
              step length= 0.2289E-03, df=-0.9948E-04,
derivative=-0.1670
              step length= 0.7901E-03, df=-0.2008E-03,
derivative=-0.1136E-03
 iter num    4, nb calls=   9, f=  1.983
 linear search: initial derivative=-0.1867
              step length= 0.5693E-03, df=-0.3642E-03,
derivative=-0.1520
              step length= 0.3055E-02, df=-0.1073E-02,
derivative= 0.1477E-02
 iter num    5, nb calls=  11, f=  1.981
 linear search: initial derivative=-0.7520E-01
              step length= 0.4256E-01, df= 0.1557E-02,
derivative= 0.1898
              step length= 0.1263E-01, df=-0.3202E-03,
derivative= 0.2478E-04
 iter num    6, nb calls=  13, f=  1.981
 linear search: initial derivative=-0.5067E-01
              step length= 0.1823E-01, df=-0.5014E-03,
derivative=-0.2876E-01
 iter num    7, nb calls=  14, f=  1.981
 linear search: initial derivative=-0.3599E-01
              step length= 0.5066E-01, df= 0.7661E-01,
derivative=  5.353
              step length= 0.5066E-02, df= 0.7205E-03,
derivative= 0.5514
              step length= 0.5066E-03, df=-0.1777E-05,
derivative= 0.2322E-01
 iter num    8, nb calls=  17, f=  1.981
 linear search: initial derivative=-0.2654E-01
              step length= 0.3260E-03, df=-0.3445E-05,
derivative=-0.2492E-01
              step length= 0.3260E-02, df=-0.2470E-04,
derivative=-0.1036E-01
 iter num    9, nb calls=  19, f=  1.981
 linear search: initial derivative=-0.1541E-01
              step length= 0.3052E-01, df=-0.4783E-04,
derivative=-0.1443E-01
              step length= 0.3052    , df=-0.3396E-03,
derivative=-0.5807E-02
 iter num   10, nb calls=  21, f=  1.980
 linear search: initial derivative=-0.1039E-01
              step length= 0.6082    , df= 0.1040E-02,
derivative= 0.4257E-01
              step length= 0.1212    , df=-0.6747E-04,
derivative= 0.3319E-04
 iter num   11, nb calls=  23, f=  1.980
 linear search: initial derivative=-0.9729E-02
```

```
                step length= 0.3506E-01, df= 0.1488E-03,
derivative= 0.3124E-01
                step length= 0.8353E-02, df=-0.1608E-04,
derivative= 0.3653E-06
 iter num    12, nb calls=  25, f=  1.980
 linear search: initial derivative=-0.8964E-02
                step length= 0.1737    , df=-0.1280E-04,
derivative= 0.1827E-02
 iter num    13, nb calls=  26, f=  1.980
 linear search: initial derivative=-0.1087E-01
                step length= 0.3524E-01, df=-0.2223E-04,
derivative=-0.8008E-02
                step length= 0.1337    , df=-0.4863E-04,
derivative=-0.1974E-04
 iter num    14, nb calls=  28, f=  1.980
 linear search: initial derivative=-0.1243E-02
                step length= 1.0000    , df=-0.9687E-06,
derivative=-0.4498E-04
 iter num    15, nb calls=  29, f=  1.980
 linear search: initial derivative=-0.2947E-03
                step length= 1.0000    , df=-0.1175E-07,
derivative= 0.1414E-05
 iter num    16, nb calls=  30, f=  1.980
 linear search: initial derivative=-0.1241E-04
                step length= 1.0000    , df=-0.2460E-09,
derivative= 0.1235E-05
 iter num    17, nb calls=  31, f=  1.980
 linear search: initial derivative=-0.1488E-05
                step length= 1.0000    , df=-0.3406E-11,
derivative= 0.7144E-08
 iter num    18, nb calls=  32, f=  1.980
 linear search: initial derivative=-0.1288E-06
                step length= 1.0000    , df=-0.2220E-14,
derivative= 0.2533E-08
 iter num    19, nb calls=  33, f=  1.980
 linear search: initial derivative=-0.6552E-08
                step length= 1.0000    , df=-0.8882E-15,
derivative=-0.2376E-09
 iter num    20, nb calls=  34, f=  1.980
 linear search: initial derivative=-0.3352E-09
                step length= 1.0000    , df= 0.1776E-14,
derivative= 0.1127E-09
                step length= 0.1000E+00, df= 0.8882E-15,
derivative=-0.2695E-09
                step length= 0.1000E-01, df= 0.6661E-15,
derivative=-0.2907E-09
                step length= 0.1000E-02, df= 0.4441E-15,
derivative=-0.3761E-09
```

```
                step length= 0.1000E-03, df= 0.1332E-14,
derivative=-0.3507E-09
                step length= 0.1000E-04, df= 0.4441E-15,
derivative=-0.3316E-09
                step length= 0.1000E-05, df= 0.4441E-15,
derivative=-0.2819E-09
                step length= 0.1000E-06, df= 0.0000E+00,
derivative=-0.3352E-09
 iter num    20, nb calls=  42, f=  1.980
***** leaves -qn code-, gradient norm=  0.3934150926394391E-08
End of optimization.

 Iteration #10 calculation time =2.266 objective function =
1.9801026
 The maximum of iterations was reached
 CREATING FLS STRUCTURE...
```

Appendix F: Heart Disease Demo

See Figures F.1–F.3.

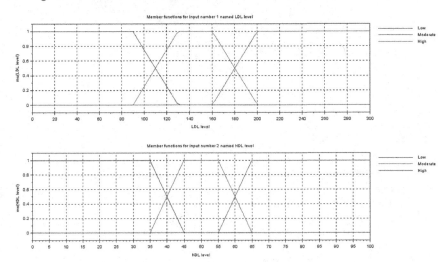

FIGURE F.1
Heart disease demo – part 1.

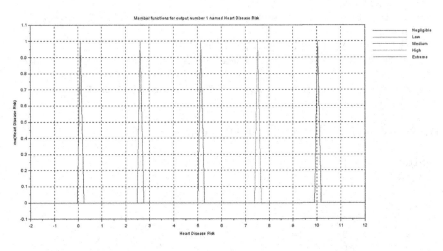

FIGURE F.2
Heart disease demo – part 2.

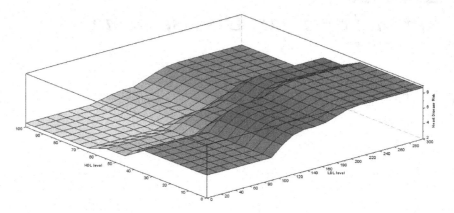

FIGURE F.3
Heart disease demo – part 3.

```
-->// Demonstrate the use of the Fuzzy Logic Toolkit to read
and evaluate a
-->// Sugeno-type FIS stored in a file.
-->// Read the FIS structure from a file.
-->// (Alternatively, to select heart_disease_risk.fis using
the dialog,
-->// replace the following line with
-->//    fis = readfis ();
-->fis = importfis(demo_path+'/heart_disease_risk.fis');
-->// Plot the input and output membership functions.
-->scf();plotvar (fis, 'input', [1 2]);
-->scf();plotvar (fis, 'output', 1);
-->// Plot the Heart Disease Risk as a function of LDL-Level
and HDL-Level.
-->scf();plotsurf (fis,[1 2],1,[0 0]);
-->// Calculate the Heart Disease Risk for 4 sets of LDL-HDL
values:
-->printf ("\nFor the following four sets of LDL-HDL
values:\n\n");
```

For the following four sets of LDL-HDL values:

```
-->ldl_hdl = [129 59; 130 60; 90 65; 205 40]
 ldl_hdl  =

    129.    59.
    130.    60.
    90.     65.
    205.    40.
-->printf ("\nThe Heart Disease Risk is:\n\n");
```

The heart disease risk is as follows:

```
-->heart_disease_risk = evalfls (ldl_hdl, fis, 1001)
 heart_disease_risk  =

    3.125
    3.125
    0.
    8.75
```

Appendix G: Linear Tip Demo, Mamdani Tip Demo, Sugeno Tip Demo

G.1 Linear Tip Demo

See Figures G.1 and G.2.

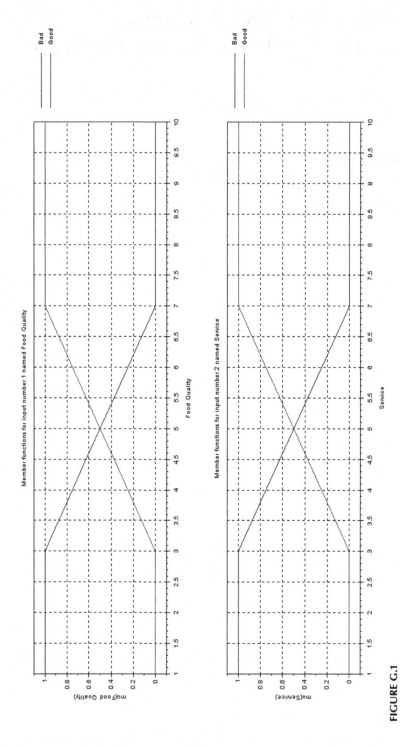

FIGURE G.1

Linear tip demo – part 1.

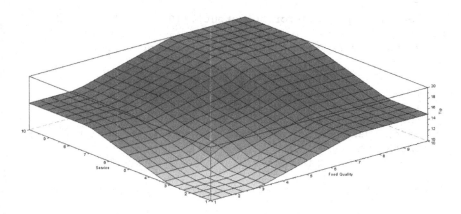

FIGURE G.2
Linear tip demo – part 2.

G.2 Mamdani Tip Demo

See Figures G.3–G.5.

```
-->// Demonstrate the use of the Octave Fuzzy Logic Toolkit to
read and evaluate a
-->//  Mamdani-type FIS stored in a file.
-->// Read the FIS structure from a file.
-->fis=importfis (demo_path+'/mamdani_tip_calculator');
-->// Plot the input and output membership functions.
-->scf();plotvar (fis, 'input', [1 2]);
-->scf();plotvar (fis, 'output', [1, 2]);
-->// Plot the Tip and Check + Tip as functions of Food-
Quality and Service.
-->scf();plotsurf (fis, [1 2], 1,[0 0]);
-->scf();plotsurf (fis, [1 2], 2, [0 0]);
-->// Calculate the Tip and Check + Tip using (Food-Quality,
Service) = (4, 6).
-->[output, rule_input, fuzzy_output] = evalfls ([4 6], fis,
1001);
-->// Plot the first aggregated fuzzy output and the first
crisp output (Tip)
-->// on one set of axes.
-->x_axis = linspace (fis.output(1).range(1), fis.output(1).
range(2), 1001);
-->figure('figure_name', 'Aggregation and Defuzzification for
Input = (4, 6)');
-->plot (x_axis, fuzzy_output(:, 1), "b", 'LineWidth', 2);
WARNING: Transposing row vector X to get compatible dimensions
-->crisp_output = mfeval(x_axis, 'constant', output(1));
```

```
-->plot ([crisp_output,crisp_output], [0 1], "r", 'LineWidth',
2);
-->legend(";Aggregated Fuzzy Output;","Crisp Output = "
string(output(1)) "%;"];
risp Output = " string(output(1)) "%;"]
                                              !--error 2
Invalid factor.
at line       56 of exec file called by :
ipt_path) then exec(script_path, -1);end;clear script_path;;if
exists("%oldg
while executing a callback
```

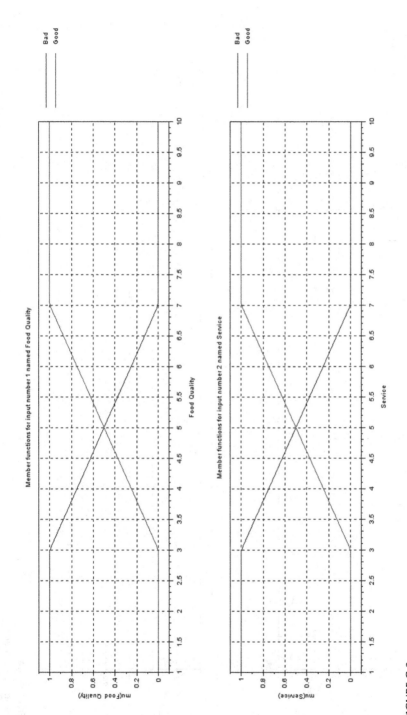

FIGURE G.3
Mamdani tip demo – part 1.

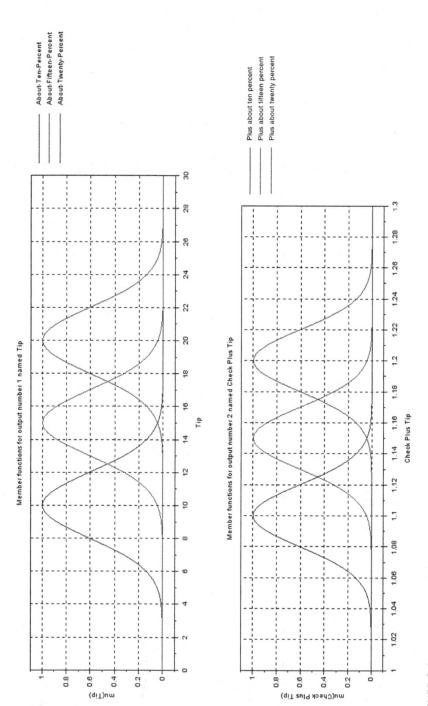

FIGURE G.4
Mamdani tip demo – part 2.

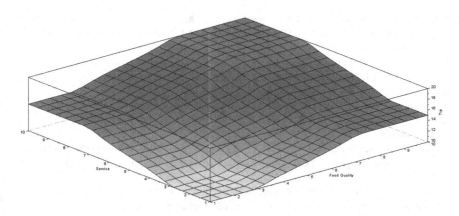

FIGURE G.5
Mamdani tip demo – part 3.

G.3 Sugeno Tip Demo

See Figures G.6–G.9.

```
-->// Demonstrate the use of the Fuzzy Logic Toolkit to read
and
-->// evaluate a Sugeno-type FIS with multiple outputs stored
in a text
-->// file. Also demonstrate the use of hedges in the FIS
rules and the
-->// Einstein product and sum as the T-norm/S-norm pair.
-->// Read the FIS structure from a file.
-->fis = importfis (demo_path+'/sugeno_tip_calculator.fis');
-->// Plot the input and output membership functions.
-->scf();plotvar (fis, 'input', [1 2]);
-->scf();plotvar (fis, 'output', [1 2 3]);
-->// Plot the cheap, average, and generous tips as a function
of
-->// Food-Quality and Service.
-->scf();plotsurf (fis, [1 2], 1,[0 0]);
-->scf();plotsurf (fis, [1 2], 2,[0 0]);
-->scf();plotsurf (fis, [1 2], 3,[0 0]);
-->// Demonstrate showrule with hedges.
-->printrule (fis);
R1: IF (Food—Quality IS  extremely Bad) AND (Service IS
extremely Bad) THEN (Cheap—Tip IS  extremely Low)(Average—Tip
IS  very Low)(Generous-Tip IS Low) weigth=1
R2: IF (Food—Quality IS Good) AND (Service IS  extremely Bad)
THEN (Cheap-Tip IS Low)(Average-Tip IS Low)(Generous—Tip IS
Medium) weigth=1
```

```
R3: IF (Food-Quality IS  very Good) AND (Service IS  very Bad)
THEN (Cheap-Tip IS Low)(Average-Tip IS Medium)(Generous-Tip IS
High) weigth=1
R4: IF (Food-Quality IS Bad) AND (Service IS Bad) THEN (Cheap-
Tip IS Low)(Average-Tip IS Low)(Generous-Tip IS Medium)
weigth=1
R5: IF (Food-Quality IS Good) AND (Service IS Bad) THEN
(Cheap-Tip IS Low)(Average-Tip IS Medium)(Generous-Tip IS
High) weigth=1
R6: IF (Food-Quality IS  extremely Good) AND (Service IS Bad)
THEN (Cheap-Tip IS Low)(Average-Tip IS Medium)(Generous-Tip IS
very High) weigth=1
R7: IF (Food-Quality IS Bad) AND (Service IS Good) THEN
(Cheap-Tip IS Low)(Average-Tip IS Medium)(Generous-Tip IS
High) weigth=1   ˙
R8: IF (Food-Quality IS Good) AND (Service IS Good) THEN
(Cheap-Tip IS Medium)(Average-Tip IS Medium)(Generous-Tip IS
very High) weigth=1
R9: IF (Food-Quality IS  very Bad) AND (Service IS  very Good)
THEN (Cheap-Tip IS Low)(Average-Tip IS Medium)(Generous-Tip IS
High) weigth=1
R10: IF (Food-Quality IS  very very Good) AND (Service IS
very very Good) THEN (Cheap-Tip IS High)(Average-Tip IS  very
High)(Generous-Tip IS  extremely High) weigth=1
-->// Calculate the Tip for 6 sets of input values:
-->printf ("\nFor the following values of (Food Quality,
Service):\n\n");
```

For the following values of (Food Quality, Service):

```
-->food_service = [1 1; 5 5; 10 10; 4 6; 6 4; 7 4]
 food_service  =

    1.      1.
    5.      5.
   10.     10.
    4.      6.
    6.      4.
    7.      4.
-->printf ("\nThe cheap, average, and generous tips
are:\n\n");
```

The cheap, average, and generous tips are as follows:

```
-->tip = evalfls (food_service, fis, 1001)
 tip  =
```

10.	10.	12.5
10.868109	13.680879	19.138418
17.5	17.5	20.
10.603913	14.20818	19.451508
10.426602	13.687078	19.032886
10.470756	14.357542	19.352533

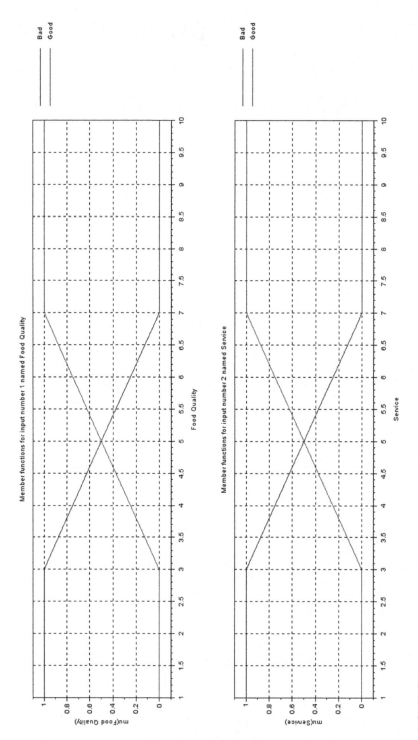

FIGURE G.6
Sugeno tip demo – part 1.

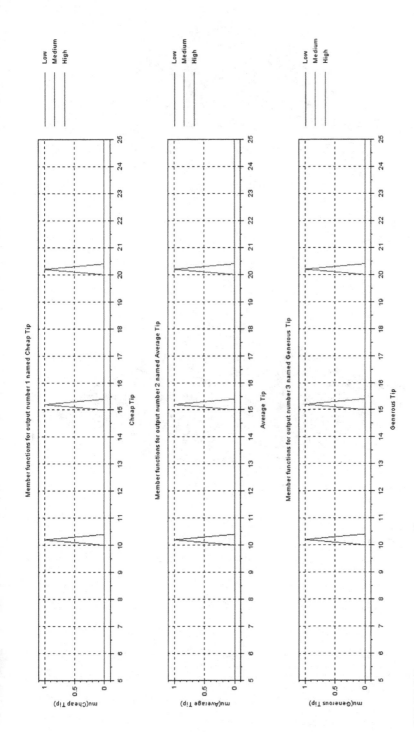

FIGURE G.7
Sugeno tip demo – part 2.

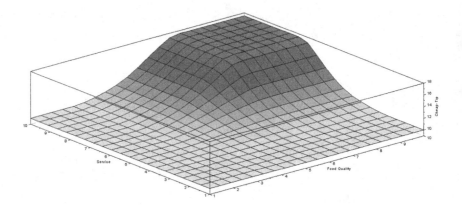

FIGURE G.8
Sugeno tip demo – part 3.

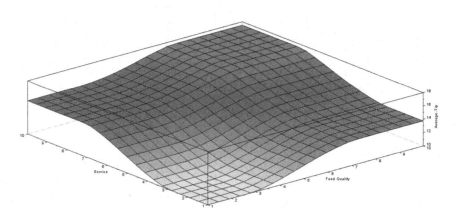

FIGURE G.9
Sugeno tip demo – part 4.

Appendix H: Miscellaneous

See Figures H.1–H.4.

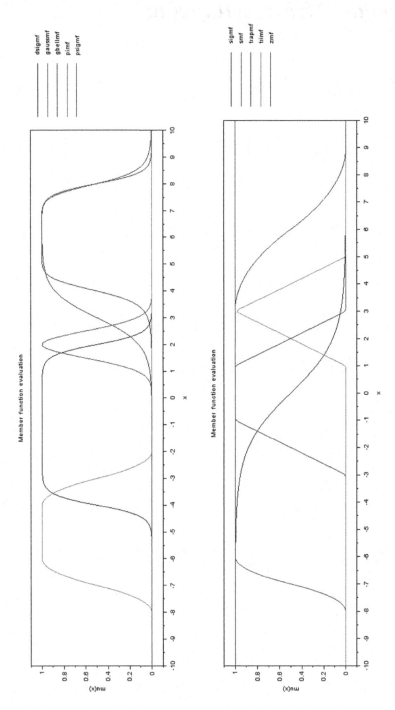

FIGURE H.1
mf gallery.

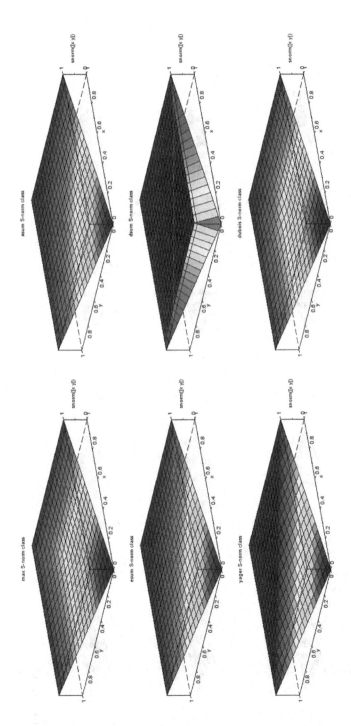

FIGURE H.2
S-Norm gallery.

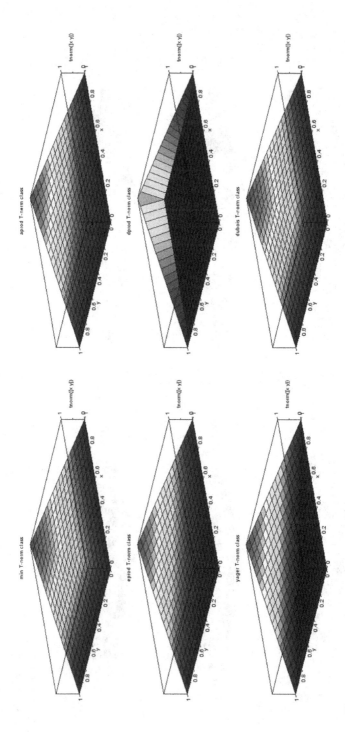

FIGURE H.3
T_Norm gallery.

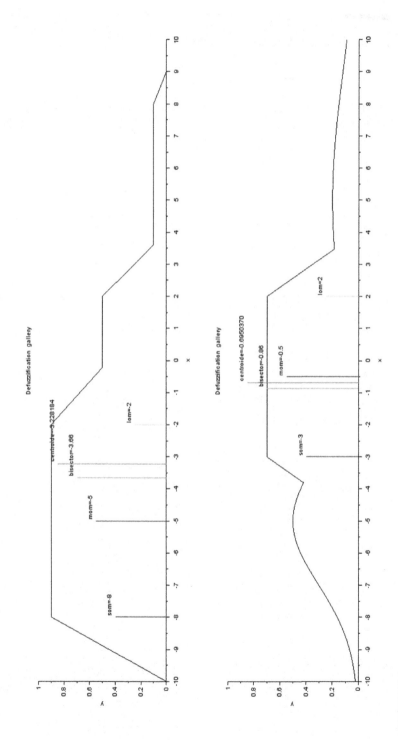

FIGURE H.4
Defuzzification gallery.

H.1 Function Fuzzy Approximation 1

```
The function to approximate is:
f(x)=sin(x).*cos(3*x)
The input domain is [-%pi %pi], the output domain is [-1 1]
Press ENTER to start.
-->
The end
```

See Figure H.5.

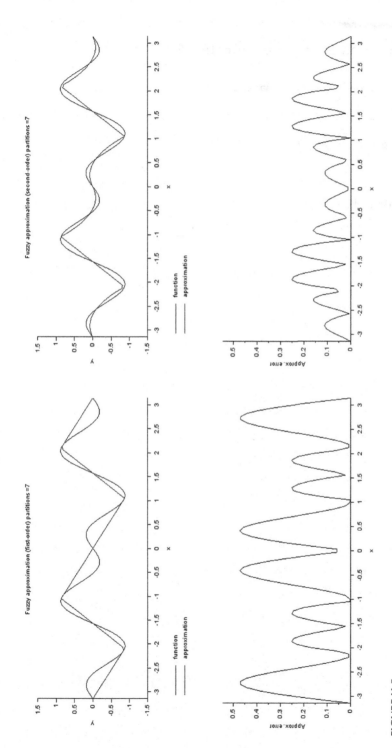

FIGURE H.5
Function fuzzy approximation 1.

H.2 Function Fuzzy Approximation 2

```
The function to approximate is:
f(x)=sin(x(:,1)).*cos(3*x(:,1))
The input domain is [-%pi %pi]x[-%pi %pi], the output domain
is [-1 1]
Press ENTER to start.
-->
The end
```

See Figures H.6–H.13.

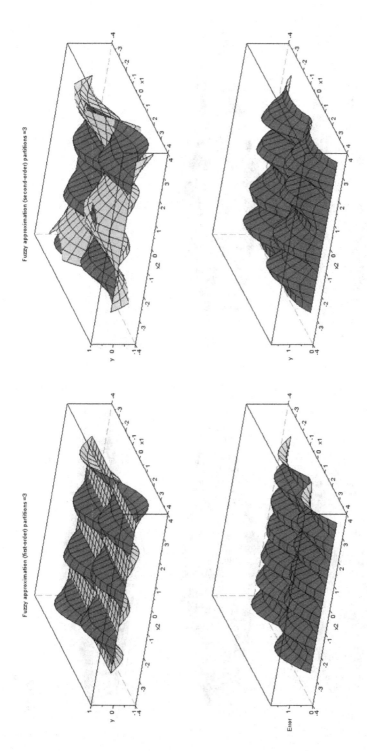

FIGURE H.6
Function fuzzy approximation 2.

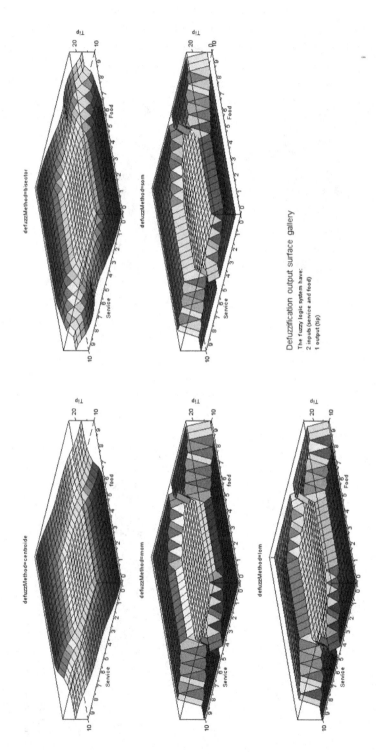

FIGURE H.7
FIS imported.

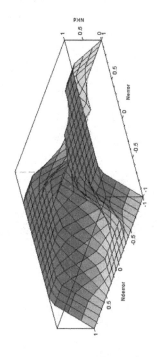

Normalized PID output surface

The fuzzy logic system have:

2 inputs (normalized error and normalized diff. error)

3 outputs (normalized Kp, normalized Kd and alpha)

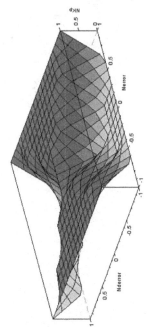

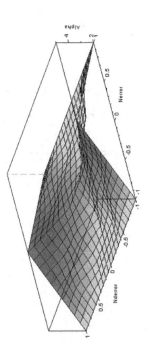

FIGURE H.8
Normalized PID.

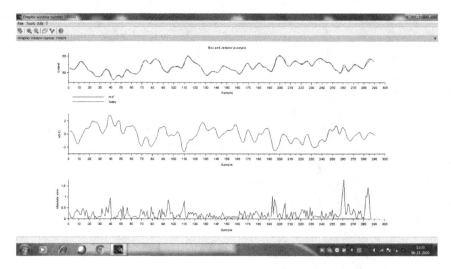

FIGURE H.9
Box and Jenkins fuzzy approximation.

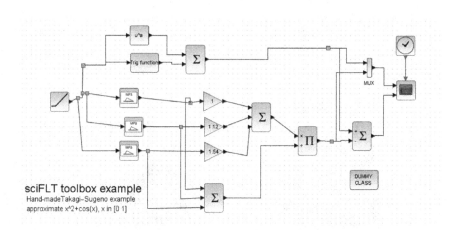

FIGURE H.10
Scicos 1.

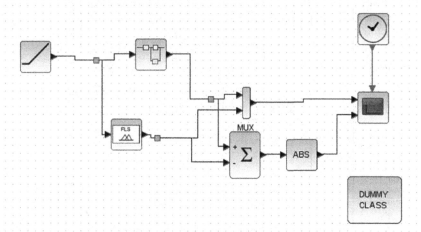

FIGURE H.11
Scicos 2.

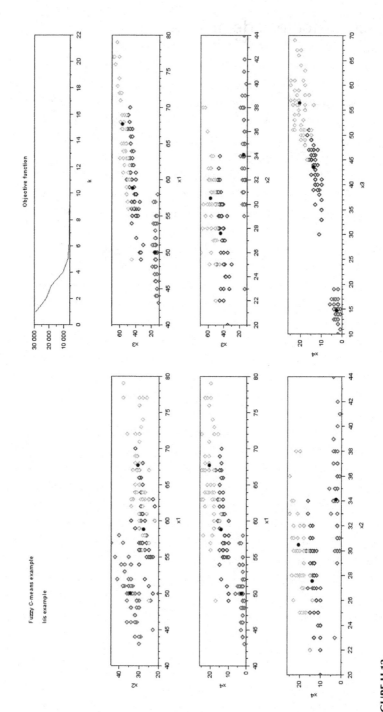

FIGURE H.12
C – means example.

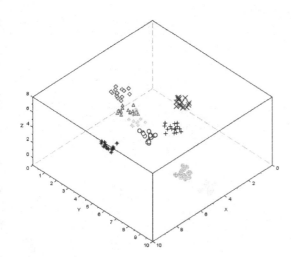

FIGURE H.13
Subtractive clustering.

Index

Note: **Bold** page numbers refer to tables and *italic* page numbers refer to figures.

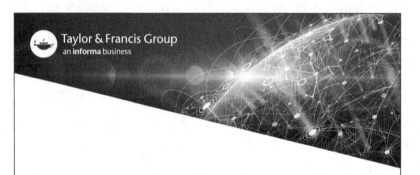

Printed in the United States
by Baker & Taylor Publisher Services